Boondocker Ballet

by
Melvin H. Thomas

DORRANCE PUBLISHING CO., INC.
PITTSBURGH, PENNSYLVANIA 15222

ISBN # 0-8059-3759-5
Printed in the United States of America

Second Printing

For information or to order additional books, please write:
Dorrance Publishing Co., Inc.
643 Smithfield Street
Pittsburgh, Pennsylvania 15222
U.S.A.

Contents

Introduction v

Glossary ix

World War II 1

    Marine Boot Camp 4

    The Pacific 8

    Tarawa 1943 17

    Mariana Islands 25

    Japan 61

Stateside and Home 1945 69

Camp Pendleton and the Pacific Crossing 92

    New Zealand Remembered 96

    Hawaii Remembered 98

    Tarawa Remembered 101

    Saipan Remembered 103

Korea 106

    Ambush on the Pyonggang 112

    A Test of Endurance 115

    A Break from the Front Lines 118

Slaughterhouse on the Soyanggang 121

The F.O. Team Wiremen 127

Retaking Lost Ground.................................... 129

Casualties Mount....................................... 132

Baptism by Fire 136

The Lost Patrol.. 141

Punch Bowl 143

Korean Marine Corps 147

B-1-11.. 152

Homeward Bound....................................... 154

Camp Lejeune, North Carolina, 1952........................ 158

Vieques, Puerto Rico and the Mediterranean Cruise............... 162

Epilogue 181

Introduction

I called my father and talked to him for some time on his eighty-seventh birthday. He said that since I was the most active member of the family that I should write down all the events and stories that had taken place during my thirty-four years of government service. He went on to say that you never know when, because of age, things will become dim and start to fade. He made me promise and I told him I would do the best I could with the limited knowledge of writing I possess. He went on to say that somewhere down the genealogy line, my children, grandchildren, and possibly my great-great-grandchildren might be interested in reading war stories of one who has lived them.

It would take two lifetimes to record the events, feats, and defeats that have occupied much of my life, but I will try and do my best to record the most interesting experiences of my service career. I don't wish to be boring in this narrative, so I'll try and be as brief as possible, and still convey the stories as I remember them. I will eliminate a lot of my reasoning and conjecturing in some phases of the stories. In writing, I will stir up memories of events and experiences that are going to be very painful—things that I have been trying consciously and subconsciously to block out of my mind. I will revive emotions that I hoped would remain buried forever. Many of them I will write about, but a few I cannot, and they will have to remain buried forever.

Do not judge anything in my stories until you've read the complete story. To those who read these stories, they will seem very daring, and at times just plain idiotic or dumb. At times it may seem that I risked my life needlessly, but in a combat area lives are in constant peril. It doesn't really matter where you are, or what you're doing, the danger is always present, so you might as well carry on and perform your duties, pursue your objectives, and ignore the fears that try to keep you in a foxhole, like a scared rat.

There is no surefire formula for survival. Most of my combat was in areas where the stench of the dead was always prevalent. You had to live with this at all times, knowing that at any moment yours might be the next body putrefying in the hot tropical sun. Each marine in this situation is only trying to live long enough to see the sun come up one more time. That is as farsighted as anyone can be in any combat zone. Our lives were measured in seconds; mine was no exception. When your world explodes, and the losses are great, you wonder why you're still alive.

The smell of the dead seeps into your clothing and the air around you, to such an extent that you can still smell the faint odor of burning, or decaying flesh, long after you have departed a combat area. On most of the islands of the Pacific, the stench of the dead vied for dominance over the fumes and vapors of diesel oil and the smell of cordite or some other type of exploding ammunition. Even now, after all these years, I can still detect the sickening odor of the dead whenever I'm around diesel oil or smoke from burning gunpowder. Not only do these smells awaken odors in my subconscious mind, they also drag up memories of buddies lying in the tropical heat covered with flies. These are some of the things I will never be able to forget nor be able to erase from my mind.

One thing that helped me to survive in combat and in other stressful situations was to keep moving at all times, even when under fire. All my life I have been hyperactive, and I feel that my brain would self-destruct if it went idle or had nothing on which to feed. Moving was one of the best defenses in overcoming this hurdle. I could always keep my mind occupied by fighting, or checking, or aiding the wounded.

I have seen men under extreme stress or pressure, fall down, crying like a baby, having to be led away by the medic or corpsman. Some of the men would display various symptoms depending on how stressful the situation was. At times I have felt close to the breaking point, but I never gave in. I would always hold on and try to make it to the next objective or mountain. Hold on there for one more day. Everyone has a breaking point. For me tenacity and determination (to a high degree) were about the only things that helped me overcome the fear of mental collapse.

On one island we had a large number of men crack up in a very short period of time. At the time we were in training and waiting for the next campaign or combat mission. The stress of thinking about the next battle was far more than some of these men could take. Fear of what lay ahead was far more demoralizing than the real confrontation.

I have never understood why some men joined the Marines. They wanted to serve but were not prepared mentally to fight. At one time we had a man taken to the field hospital with mental, or battle, fatigue. I asked an old sergeant what he thought caused the man to break down. He replied, "All of us are crazy to some extent; some just show it more than others."

I didn't know what patriotism meant when I joined the marines. I joined for travel, adventure, and the desire to be part of a marine fighting unit. All of these things would be realized before my career in the marines would come to a sudden halt.

I will relate a number of things that are still very painful. It will awaken rage, hate, and anger at things that have taken place. Things that could have been prevented, if only this or that had been done. It's easy to judge events and things after the fact. I don't intend to do this. I will state them, maybe remark about some of the occurrences, but keep my praise and anger to a minimum.

For years, and right up to the present, I still have nightmares or bad dreams. They don't occur as often as they used to, and I have become adjusted to them, as they always seem to end in the same way and have moderated with time.

My years in the marines are like a plane leaving a vapor trail. I can always go forward, but the past, the men, the battles, the highs, and the lows, all of the things that have accumulated to make and shape my life are fading fast. If anything is to be recorded of the existence of my life on earth, it will have to be now.

I wish to say, "This is not a history of wars, governments, the Marine Corps, or marine units. It's my life, a life I was cast into by wars, and how I lived, survived, and served to overcome the ordeals and obstacles that happened to confront me along the way."

I have been very fortunate in knowing and serving with some of the most courageous, gutsy, and gung-ho marines who ever planted a boon-docker on foreign soil. Boondockers are the marines' rawhide field shoes. When you run a gauntlet of fire, whether it be artillery, mortars, snipers, or a platoon of riflemen, you are high-stepping, slipping, sliding, and dodging behind anything to make a difficult target for the enemy. I call this doing the boondocker ballet.

This is my personal autobiography. It's taken me quite a while to complete this narrative. I have put it aside many times, only to come back, pick the pen up again, and continue to record the trials and successes of the past.

I sincerely hope that anyone who reads my story might benefit by reliving and experiencing some of the events and escapades of my life.

I dedicate this book to the marines with whom I was most closely associated during wartime—the 75-mm pack howitzer unit of F-2-10, the Sixth and Eighth Marines of the Second Marine Division in the Pacific during World War II; and in Korea, the men of F-2-5 and B-1-11 of the First

Marine Division. For years these men were the only family I knew. They have made an indelible imprint on my life. The memory of some of these men, both living and dead, will always be with me. I could never have chosen a better group of fighting men to serve beside than the ones I have just mentioned.

Glossary

I've eliminated most of the Marine Corps slang to aid any non-marine in reading my autobiography, but I've included a list of some of these words for your information.

BAMS	Broad Ass Marines
Boondocker	Rawhide field shoes
Boondocks	In the field or some other God-forsaken place
Bulkhead	Wall
Civvies	Not a uniform
Chick	A girl or the youngest-looking Marine in the unit
Chow	Food
Deck	Floor
Deep Six	Throw away, file 13, or shit-can it
Doggie or Dogface	United States Soldier
Foxhole	Depression in the ground generally shared with a buddy
Gentleman	A Marine officer, but said with a sneer
Gook	Native west of California

Head	Toilet or washroom
Hit the Deck	Turn loose your cock and grab your sock
Horse Cock	Lunch meat, shaped as
Housewife	Sewing kit
In the Field	See Boondocks
Liberty	Away from your duty station
Ninety-day Wonder	Newly made second lieutenant
Bellhop	Hotel employee
Red lead	Catsup
Seconds	More of anything
Piss Cutter	Overseas cap
Scuttlebutt	Drinking fountain, lister bag, or rumor-gossip
782 Gear	Web equipment, carriage belt, canteen, etc.
Short Timer	Nearing discharge or transfer
Skipper	The CO or the old man
Skivvies	Drawers, knee, cotton, and T-shirt
Soyanggang, Pyonggang	Two small rivers in central Korea
Split Tails	Female
Swabbie	Sailor deckhand, not corpsman

World War II

It was late on a Sunday evening when my brother and I came out of the Texas Theatre on West Jefferson Avenue in the Oakcliff section of Dallas. This was the first time we had heard of the bombing of Pearl Harbor. Neither one of us had the faintest idea where Pearl Harbor was located. It wasn't until after the president declared war on the Japanese that we learned where Pearl Harbor and Hawaii were located on the map.

I was sixteen years old at the time, and Bernnie was seventeen. I had hitched a ride from the South Texas town of Cuero to spend a weekend at home. I was in the National Youth Administration studying to be an electric welder. After my visit to Dallas I returned to Cuero to continue my training. A month later I was transferred to the NYA in South Houston. This NYA would try to place boys in jobs once they had completed their training. I didn't have much hope of being placed in a job, since I was under the age of eighteen. While training and waiting, I saw most of my close friends leave. Some went home, some obtained work in the oil fields, and others found work in the Houston shipyards.

One morning I asked the administrator if I could leave and try to get a job on my own. I was advised that I could live in the camp until I became of age, but if I left on my own for any reason it would be for good. I would not be allowed to return. I packed my old suitcase and left the following morning with only fifty cents in my pocket. I hitched a ride to Pasadena, Texas, and used a dime to check my suitcase in a locker at the Greyhound bus station. I then caught a ride to the Deep Water Inn. This was the turnoff to the Houston shipyard. I was both disappointed and amazed on entering the shipyard. The employment line was over 300 feet long. I got in line but had little hope of getting on as a welder. A large sign above the employment window read, "Only first class welders with equipment need apply."

1

I decided to remain in line and try to get on as a laborer or whatever job might be available. It was late in the evening when I finally reached the employment window. The only job opening left was for a ship fitter's helper. I filled out the application, was interviewed for the job, and told to report to work the following morning. I listed my age as eighteen. I didn't have my birth certificate with me, but the interviewer told me I had six weeks in which to submit it to the company.

I got back to Pasadena at about five in the evening, and started making the rounds of the rooming and boarding houses. I was trying to find one that would extend credit until I received my first paycheck. Pasadena was a very small town, and I had made most of the rooming houses by nine o'clock. Credit seemed to be a bad word to these people; none of them would let me stay on credit.

I retrieved my suitcase from the bus station, drank a cup of coffee, and walked to the edge of town to try and hitch a ride to Dallas. I spent over an hour on the edge of town, but no one offered me a ride. The traffic was very slow. Behind me was a big neon sign that read, "Womacks Boarding House." I had nothing to lose, so I decided to try one more time. It had been over twenty-four hours since I had eaten, so I was getting desperate. A lady answered my knock and I explained my plight or predicament. The lady asked me in, took me to the kitchen, gave me a plate of food, and after I had eaten, showed me to a room. I was tired, so as soon as I showered I went to bed. I was awakened early the next morning with everyone getting ready for work. After breakfast I was given a sack lunch to take to work. One of the men gave me a ride to the shipyard. These people were really nice to me, and I stayed with them for quite some time. The only complaint I had was the awful smell of the paper mill. Some said you would get used to the odor, but I never did.

After six months, I was promoted to fourth class ship fitter.

I learned where the welding superintendent's office was located and gave him a visit on my lunch period. He was very cordial, and we had a long talk. The following morning I went to work as a second class welder. One month later I took the test for first class welder and passed. This was a good paying job, and I could work Saturdays and Sundays for overtime. It was at this point that my employers discovered I had not given them my birth certificate. I had received it from home and told them I would bring it in the next morning.

I was sure that this would be the end of my job, so I got my things in order and prepared to go back to Dallas. The following morning I turned the certificate in to the foreman. I expected to get a final notice at any time, but none came. My birth certificate was clipped to my next paycheck. I made a remark about this at the supper table, and one of the boarders said there was no way anyone under eighteen would be allowed to work in the shipyard. He asked if he could see my birth certificate. After looking it over he asked, "What year were you born?" I said, "I was born on March 24, 1925." The boarder told me the date was wrong. It showed I was born on March 24, 1924. This would make me eighteen. It saved my job, but then I should have already registered for the draft. Bernnie, my brother, was born in 1924. He was eighteen and had just enlisted in the marines. I continued to work at the shipyard. I was getting good pay and didn't want to give it up.

Everyone in the boarding house was worried about me, and kept after me to go and register for the draft. They were afraid I would end up in jail if I didn't comply with the law. I waited as long as I could, but finally gave in and went to the draft board. I told them that there was a mistake on my birth certificate. I really was only seventeen. They gave me a knowing look, and told me to go ahead and register, that it would probably be three months or more before I was called. What a laugh! Three days after I registered for the draft I was given a physical and passed. I selected the marines as the branch of service I wanted to enter. The marines gave me another physical, which I passed. Ten days from the time I had registered for the draft I was in San Antonio being sworn into the marines. A number of new recruits and I boarded a Pullman train and spent most of a long and tiresome three days traveling to San Diego for boot camp training.

Marine Boot Camp

Upon our arrival in San Diego, a big marine drill sergeant met us on the docks of the train depot and did his best to march us in an orderly manner to the waiting bus. When we arrived at the marine recruit depot, everyone was yelling at us, "You'll be sorry." We debarked with our suitcases and stumbled into what was supposed to be an orderly formation, with laughter and jeers from the older boots.

I won't dwell too much on boot camp training, this has already been hashed and rehashed so many times in the media that the rough training is common knowledge. I can assure you, it was anything but a picnic.

We were assigned to six-man tents, up at five-thirty every morning for roll call. Then we made up our bunks, rolled the tent flaps up and, using hand brushes and buckets, scrubbed the tent decks with soap and water. The mornings in San Diego were very cold, and our hands became numb during this morning routine, but that was carried out every morning without fail. By the time we went to morning mess I was so cold that a cup of coffee was a life-saver. This would be our home for the next seven weeks. There would be no PX privileges during the first six weeks of training.

About five percent of our platoon was dropped for various reasons. Those men were sent to the casual company and would later be sent home. Being washed out went hard for some of these men; they had had their hearts set on being marines. It would be degrading for some of them to return home after having had a big send-off party given by friends and families.

I didn't have a party. I joined the marines in Houston, and I didn't even get to go home for a visit, much less a party. It had been a long time since I had been home; it would be a few more years before I would see the family again. One thing surprised me about boot camp. There was fresh milk on the tables, and you could drink all you wanted. This was a precious commodity back home. We always had so many small ones that we left the milk for them.

Boot camp was very hectic, to say the least. I was always edgy and would jump at the first command. One morning while in the Camp Matthew rifle range someone yelled, "Fall out." I immediately and, without thinking, jumped up, ran outside, and got into formation. The NCO in charge marched us up to a building and gave the command, "File in." I found myself in the middle of a Catholic church service. I had forgotten that it was Sunday. I sat through the full service. I don't suppose it hurt me, but I was on full alert the following Sunday.

My brother, Bernnie, came to visit once while I was in Camp Matthew. He was in training at Camp Pendleton. He told me what I had to look forward to, and how to conduct myself to finish the rest of boot training. He said the worst was over, and with his encouragement I was determined to complete the course.

On entering boot training, we were told to shave every morning. Anyone caught with fuzz on his face would be punished. I completed the full seven weeks and never once shaved. I didn't have a razor or a beard.

Graduating from boot camp was a grand occasion with the marine band playing marches during the pass and review parade. Many friends and relatives came to see the graduation ceremonies. Campo was my best boot buddy, and like me had no one come to visit on graduation day. It was a proud day; every man felt a sense of pride in accomplishing his first big mission in the marines. It had been a tough seven weeks. I can never forget the men in Platoon 220, who suffered with me through the rough training and treatment we had undergone in such a very short time. We had learned a lot; we were no longer boots but full-fledged marines. We were now qualified to wear the Marine Corps emblems on our lapels.

My buddy, Campo, was from Detroit, Michigan. We were hoping that we would stay together when we left boot camp. This was not to be. Campo was sent to Camp Pendleton, and I was assigned to amphibious tractor training in Boat Basin, a small camp about three miles north of Oceanside, California. There was nothing more than six barracks and a long boat dock.

Amphibious tractors were called water buffaloes, or floating coffins. I enjoyed driving them out into the water, and then bringing them back in through the foaming surf. Most of the time about half of them developed some kind of trouble, and we would spend most of the day towing the dead ones back to shore. Besides the operational part of the training, we had to attend mechanics class. I hated this class. I couldn't stay awake. After being caught dozing a few times, I was put on report. The CO wanted to know why I couldn't stay awake. I told him that I was bored, that I had joined the marines to fight, not to be a mechanic. The next day I was transferred to Camp Pendleton. I was assigned to a combat conditioning course. This course was to prove much rougher than infantry or boot camp training in many ways. I enjoyed this training with forced marches, knife and bayonet fighting, hand-to-hand combat mixed with Judo, and of course, self-discipline. We also went through survival training and many night exercises. We made a number of twenty-mile hikes with only one canteen of water. Many times I limped along with blisters all over my feet. I finally learned to soak my boondockers in a bucket of water for twenty-four hours, put them on wet, and wear them until they were dry. This insured a perfect fit, as the wet rawhide of the shoes would mold itself around my feet. I could never get shoes that fit perfectly, so I used this process as long as I was in the Marine Corps. This course also included the art of snapshooting. This was shooting at silhouette targets without aiming. I became rather good at this, and fired the final course with a score of nineteen out of a possible twenty. Just about everything I learned in that course has helped to save my life at one time or another, either in the military or in civilian life.

Liberty in Camp Pendleton wasn't very good. I only knew one road in and out of that camp. That road was always clogged with marines trying to hitch a ride. Most of the time I would bypass them and hike the eighteen miles to the main gate. I always wore boondockers which were spit-shined to a high-gloss polish. They looked as good as dress shoes—no one could tell them apart unless they looked closely. Once I made the main gate, I would hitch a ride to Long Beach or Los Angeles. I made about seven liberties in Los Angeles and Long Beach. I visited the Hollywood Canteen once, where Hollywood stars often volunteered to work. I looked around, had a cup of coffee, but never saw a movie star.

I met a pretty Jewish girl in Hollywood, and had a couple of dates with her, dancing at the Palladium to the bands of Tommy Dorsey and Johnny Long. This was a platonic relationship, so I'll give her name—Gladys Ponessa. I'll mention her again later in this book. After a date with her, I would put her on a bus to Burbank and then catch a bus back to Los Angeles and either catch a bus or hitchhike a ride back to Oceanside. I would then walk the eighteen miles back to the barracks, arriving just in time for the morning roll call.

While on liberty in Los Angeles, I met the most beautiful brunette I have ever seen. We attended a movie together. On our next date, I met her in the lobby of a hotel in Long Beach. The first date had been very good; I could hardly wait to see her again. She was a very pretty and sweet girl. It would be easy to fall in love with this girl. I thought I had made a good impression on her, but on this date she asked me to call a girlfriend

6

of hers, and we could make it a threesome. This came as a shock to me, and was the first time I realized she was a lesbian; she wanted to have a lesbian relationship with the girl she wanted me to call. I realized she was using me, and I refused to make the call and told her I was leaving. This ended the relationship; it was a hard choice to make because she was so pretty. A few years later I saw the picture and the story in a detective magazine about the murder of the Black Dahlia. I was amazed at the resemblance to the girl I had dated. I clipped a picture of her out of the magazine and carried it for years. I've read everything I can find on the Black Dahlia's murder. In all of the stories I've read, including the police investigation, I've never come across one thing that indicated the Black Dahlia was ever involved in a lesbian relationship. It's possible that I dated another girl, but I still wonder about this after all these years. To this day, I have never been with a prettier girl.

I completed my training at Camp Pendleton, and received additional training at Camp Elliot.

Bernnie had left for overseas with the Eighteenth Replacement Battalion over a month earlier. I was right behind him, being assigned to the 22nd Replacement Battalion.

I mailed my APO card home so the folks would know I had left the States and would know where to write.

We had been training to fight the enemy; now at last our training was over. We would soon be in the wide Pacific in the enemy neighborhood. We had trained hard, and everyone was anxious to be on the way.

The Pacific

We boarded an old Dutch freighter, the *Bloen Fontaine Craven Hagen*, in San Diego. There was a huge crowd on the docks to see us off; most were relatives waving their farewells. I didn't have any relatives in the crowd, but I waved with the rest of the men as we raised anchor and slowly got underway. I remained on deck watching the land fade away to nothing.

This was a very old and a very dirty ship. I think it had been used to haul swine or fertilizer. Things aboard this ship were bad from the start and continued to go downhill from there. Wooden toilets were hung onto the side of the ship. I was afraid to use them for fear of falling into the sea. We only had two meals a day, and both the quality and the quantity of food was inadequate.

After two weeks aboard this monster, we dropped anchor in the clear blue waters of Tongatabu, located in the Friendly Islands. This was the same Tonga mentioned by Captain Bligh of the *Bounty* many years ago. The natives came out to the ship in outrigger canoes to barter and trade with the marines. Many of the men traded playing cards for coral and sea-shell beads.

This was a very small island, and no one was allowed on liberty. The natives came aboard and put on a good show for us. The men from the island performed a daring and dangerous dance with large sticks and machetes. The timing had to be perfect, or some of the dancers could have been injured. Four of the native girls sang and danced. I thought they were rather good. Anyway it was a break from the ship's boring routine.

The enticing combination of pretty girls and crystal clear blue water was too much for two marines. They jumped ship to swim ashore. Distance can be deceiving when anchored offshore. It mistakenly appeared that we were anchored only a short distance from the island, but we were actually anchored over 600 yards from the shoreline. One of the marines drowned while swimming to shore, and the other one was declared missing. They still had not found him prior to our departure. I never heard any more about this, so I don't know what happened to the other marine.

Crossing the equator can be a dirty and messy experience. This is where the old shellbacks initiate the pollywogs into the realm of King Neptune. I was thinking ahead on this; I needed a place to hide until it was all over. The old salts would send gangs all over the ship to catch any man who had never crossed the equator before and drag them out to the main deck, butcher their hair, and then smear a mixture of hair, syrup, and eggs all over their body. There was only salt water from the ocean to shower in, and it was hard to wash this concoction off without salt water soap. We didn't have any, and the ship didn't furnish it.

When the time came, I laid down in the bottom bunk, raised and secured the outside of the bunk to the bunk above, thus concealing me. I remained there for most of the day. When it was all over, everyone asked why the shellbacks had not cropped my hair. I told them I was on mess duty. This seemed to satisfy their curiosity.

Our next stop was Noumea, New Caledonia. I hoped we would take on fresh supplies. Our morning meal was boiled rice and black coffee— nothing else, no milk or sugar. Our evening meal consisted of boiled red cabbage or boiled cauliflower with no seasoning or bread. I couldn't eat much of the food and dropped from 185 pounds down to 160 in about twenty days.

We debarked in Noumea and boarded trucks that took us out in the boondocks to the Marine Raider training area. It was dark when we reached the camp, and a cold downpour started. I stood in line in a cold driving rain, soaked, and ankle deep in mud just to get some C rations to eat. I was starving. I even went back for seconds.

This camp was a mud hole; there was mud everywhere. It had been raining intermittently for over two weeks. Mud was even ankle deep inside the tents, and the mosquitoes were big and thirsty. I woke up once during the night, and the bottom of my feet were driving me crazy from all the mosquito bites where my feet had touched the mosquito netting. From that time on I always wore my socks while in a tropical zone.

The Marine Raiders were an elite fighting force. Besides myself, many others volunteered for the Raiders. Only a few were selected; these were selected alphabetically. They never got to the *T*s, so I wasn't among those chosen. The men not selected were taken back to Noumea, and again boarded the slave ship, as we called it.

The ship had taken on supplies. You could tell by the delicious odors that drifted down from the officers' mess. However, we continued eating the same garbage we had been getting for almost a month. One raw potato came falling down the stairway from the officers' galley, and two marines had a good fight to see who got the potato.

Plowing through the South Pacific can be very hazardous for more than one reason. We were always on the alert for Japanese submarines, and strictly observed the blackout after sundown. We stopped and took on a battalion of New Zealand troops from Norfolk Island who were bound for Wellington, New Zealand.

When we were about halfway between Norfolk and Wellington we were hit by a gigantic storm. It was so fierce that no one was allowed above deck. At times the ship rolled all the way over to one side, and everyone held their breath and braced themselves to see if it would keep going or if the ship would right itself. There were many scary minutes. This went on all night, and I was relieved when it was all over. I couldn't believe a storm could be so strong as to toss a giant freighter around like a matchstick. Large waves would cap over the top of the ship with a loud boom, vibrating the ship's structure so badly I thought it would split in two.

Cruising into the harbor of Wellington was a welcome sight. Houses built on the sides of hills overlooking the city looked so inviting along with the green and lush vegetation of the countryside.

I was happy to see the city of Wellington, New Zealand. At last we would be free of this slave ship. If our government paid these people to transport us, they must have made enough to retire on for the rest of the war, just on what they saved by starving the marines. After thirty-one days aboard this piece of junk, everyone was elated to get off and not look back.

We debarked and boarded trucks to take us to the different marine units. At each camp a number of names were called, and those men would get off with their equipment. The trucks then continued to the next camp, and the same routine was repeated. We only had a few men left in our truck when we arrived in Camp Paekakariki.

I was assigned to F-2-10, an artillery battery. The battalion commander was Col. George R. E. Shell. The battery CO was Capt. Johnson. I ended up in Fox Battery all alone; I didn't know a soul. I was assigned to a tent and put in the Third Gun Section. I didn't sleep very well the first night, and was up very early the following morning. I was walking around the camp when a lone marine came toward me. Just as he got even with me, he said, "Welcome aboard, Thomas." This astounded me and caught me off guard. I didn't think anyone in this camp knew me. I found out later that this marine was John Paul Young, the youngest gunnery sergeant in the Marine Corps. He made it a point to learn the names of new men and to greet them by name. It sure made me feel better just knowing someone knew I was alive. Gunny Young was unable to grow a mustache, and wouldn't let anyone else grow one. I didn't have to worry because I still had not shaved since I had joined the marines.

The guns of Fox Battery were the 75-mm pack howitzer. I was really surprised when I went to the first gun drill. I didn't know such guns existed. My section chief was William S. Prosser, an excellent all-around marine. He had been on tour in Iceland, and had just returned from the fighting on Guadalcanal. The battery was now in training for the next campaign.

Sgt. Prosser and Sgt. Wyatt did their best to try and teach me something about artillery, but the artillery terminology they were using was beyond my comprehension. It was like a foreign language to me. I was beginning to learn, but it was a slow process. All the positions in the gun section had been assigned to the older men. Each man had a specific duty, such as gunner, ammo man, et cetera. If I learned something I didn't forget it, but I was green and inexperienced, and I didn't have the faintest idea what the word *responsibility* meant. Everyone was afraid to trust me with anything that had to do with firing live ammunition. Sgt. Prosser recognized this and never gave me a regular position in the gun section. I generally was put on working parties or relieving a man on the gun so that I could get some experience in the various positions.

Sgt. Wyatt was from Hughes Spring, Texas. Everyone called him by a nickname. While in the field one day I tried to get his attention. I thought I remembered his nickname and yelled out the name. He turned and walked up to me and asked, "What did you call me?" I said, "Bottle Ass." He got red in the face, very angry, and loud. He told me in no uncertain terms that I was not to call him anything but Sgt. Wyatt. When he left some of the men came over and asked me what had happened. I said that I thought I had called him by his nickname and he had gotten mad about it. I told them that I had called him Bottle Ass like everyone else. "Oh no," they laughed, "it's not Bottle Ass, it's Jug Butt." As long as I was in Fox Battery I never called him anything but Sgt. Wyatt. I never trusted myself with nicknames after this incident and generally called NCOs by their names and ranks.

I participated in the Foxton maneuvers which was a long march for the infantry. This maneuver was anything but fun. It rained the entire time we were in Foxton. For three days, if I wasn't doing a job or standing watch, I was huddled under a dirty oily tarp trying to keep out of the cold rain and trying to stay warm. Everyone in the battery was miserable. This maneuver was depicted in the movie *Battle Cry*, which was released after the war was over.

We made another maneuver in Waiouru, New Zealand. This was the coldest place I had ever been in up to this time. We stayed in tiny two-man huts. To wash up you had to go to an open washroom. The water was ice cold. I was thankful I didn't have shaving to worry about.

The ground around Waiouru was always frozen so we couldn't dig the guns in as we generally did. All of the time we were on this maneuver a big snow-covered mountain loomed in the background. This only added to the bleakness of the terrain. We spent most of the day getting the guns up, and through, the foothills of the huge mountain. It was here I got to see the guns fire live ammunition for the first time—not only indirect fire, but direct fire. Most of this was done in freezing drizzle. After dark the temperature would drop very fast. I would hang up my wet woolen overcoat, and the next morning it would stand up by itself, frozen solid. I would hate to live in this area. It was too bleak, drab, and always extremely cold. My big toe and the one next to it suffered frostbite but I didn't report it because I was afraid they would be amputated. It was at least three months before I started getting feeling back into them. I was very happy when we departed this arctic-like atmosphere.

The first time we set out through the foothills of the mountain going on field maneuvers I rode in the back of a 6x6 truck. We were towing the guns behind. We started winding our way up these steep and heavily wooded hills when we passed a sign that said, "Beware of the Lorries." We passed two more of the same signs along the road. I became alarmed but said nothing. We had not been issued any small arms ammo, so all I could do was keep my bayonet handy. It seems the lorries were causing a lot of trouble. I was very alert, and had my eyes scanning the terrain for any sign of the lorries. I thought the road signs were referring to a group, or tribe of natives, like the Maoris of New Zealand. It was much later when I learned that the lorries being referred to on the signs were trucks. I never told anyone about this for fear of being ridiculed. I kept mostly to myself and never asked many questions. I was very lucky; I had survived my first campaign against the lorries of New Zealand.

On a very cold day we boarded ship and got underway with all of our guns and equipment. This was to be an amphibious maneuver in Hawkes Bay. The bay was extremely cold and windy when we boarded the small boat with our gun to go ashore. I hoped our boat would land high upon the beach, so that we might have a dry landing. This was not to be; when the ramp was dropped we stepped off into waist-deep water. It took quite a while, and a lot of work, but we finally got the gun ashore. I was soaking wet and about to freeze. I had to keep moving. I was certain I would freeze to death before I ever got my clothes dried. I was as numb as a zombie.

This maneuver lasted for three days, then all the units packed up and departed. Two other men and I were left on the beach. We were left behind to guard a large ration dump. The only evidence that the marines left was the ration dump and the hulk of one tank that became stuck in the sand during the landings. We stood on the beach and watched as the last trace of a ship faded over the horizon.

The days were very cold with the wind blowing across the water. We had only one blanket and our rubber poncho for cover. During the nights the wind would pick up and bring the temperature down. We made a small tunnel with boxes of rations. This was the only windbreak available.

Late one evening, just at sundown, a bunch of dirty Maori cowboys rode up to the ration dump. They looked mean and were filthy dirty from working on the range. They surrounded us, and I thought we were going to have trouble, but they were just curious and wanted to talk. They asked us our names and where we were from. When I said, "Texas," it rang a bell; they had all heard of Texas. They were laughing, and one of them brought me a bareback horse to ride. I had never been on a bareback horse in my life. This was a huge animal, and I didn't want to ride it However they wouldn't take no for an answer, and since we were outnumbered I decided to humor them. I had no sooner straddled the horse's back when one of the Maori slapped the horse's rump. I made a 360-degree flip, landing on my butt in the sand. I took one of the saddle horses and rode down the beach to retrieve the animal. The Maori cowboys were still laughing when I returned. They were having a good time. We gave each of them a can of fruit cocktail and said goodbye; they were in good spirits as they rode off. You could still hear them laughing as they rode out-of-sight far down the beach. We remained on this cold, wind-swept, miserable beach for ten long days and nights until someone finally missed us. I'm not sure how, but someone notified the local constabulary, and they put us on a train back to the town of Paekakariki. We were as dirty as the Maori, and I'm sure we looked like fugitives from a refugee camp. We left the ration dump still stacked on the beach where we had been guarding it. The dump was now up for grabs. I hoped the Maori cowboys would get to it first.

I liked the people of New Zealand. They treated the marines with consideration. When I went on liberty I would usually buy steak and salad because the prices were very reasonable. The steak, of course, was mutton. Mutton was served in almost all of the restaurants; it was about the only steak you could get. It was also the only meat being served in our mess. The marine cooks could not prepare the mutton as well as the cooks in the civilian restaurants, or maybe the meal just tasted better when off the base.

Many times in the evenings, if I didn't have anything to do, I would go down to our galley and watch the two butchers cut up the mutton carcasses for the next day's meal. The two butchers kept a dialogue going between them as to how best to dissect the various parts. One man would make a cut, only to be berated by the other one for not doing it correctly. One of the butchers looked to be Italian, with an unmistakable Brooklyn accent, which added to the novelty of the show. It was a good show, and I went to watch every time they had meat to cut up. I was not the only

13

one to find this amusing. After two weeks you had to get there early to get a place to sit.

One day while at the base PX I had a big surprise; I ran into Bernnie. Up to this time, I had had no idea where he was, or what unit he was in. He was located only a short distance from McKay's Crossing and the main gate to this base. We were able to make a few visits. He was in the machine gun section of B-1-2. I had hoped we could go on liberty together, but for one reason or another we never did. Bernnie had two good friends in his machine gun section whom I liked very much. Their names were Jones and Taylor; I don't recall their first names.

I can't remember why, but I never made liberty in Wellington like the other men. When I had liberty, I boarded a train at McKay's Crossing and went to the town of Palmerston North. This train looked just like the ones you see in old Western movies. If the seats were full, the conductors would let the marines ride in the baggage car where you could sit on crates or boxes and watch the countryside as the train whirred by. Every time I arrived in Palmerston North I would buy two meat pies from the vendor at the train station. These were the best meat pies I have ever eaten anywhere, and of course, they were made with mutton.

The town closed down at six in the evening. If you were drinking beer in a pub, the tavernkeeper would put you out promptly and close the doors at six o'clock sharp. On Saturday nights the only thing open in Palmerston North was the A&A dance which was always crowded with marines and New Zealand soldiers. The women outnumbered the men, and most of the women were on the good-looking side with fair skin, probably due to the cold weather in this country.

I met a very pretty girl at this one particular dance; we got along perfectly. It was love at first sight for both of us, but we had to sneak around to see each other, or only meet at the dance. Her father didn't like marines and would probably have killed us both, had he caught us together. She was seventeen, sweet, and very pretty. We both knew our love could never last, but we decided to see each other anyway, and enjoy the love while, and where, we could. On one liberty, I didn't have much money, and as usual we met secretly and went for a long walk in the country. That same evening a severe storm struck unexpectedly. There was only one shelter in sight, and we had to cross a fence and make a wild dash for the building. This was a small barn, and the rain on the roof made a deafening noise. The rain, with the smell of the hay, and the musky smell of her hair were conducive to making love. It didn't escape up. We spent the whole evening huddled together in the small barn. This was one of the most cherished memories I will always have of New Zealand.

The last time I was in Palmerston North, I was on a seventy-two-hour pass. I rented a room for the weekend, and after the usual Saturday night dance, we retired to the room to spend some time together. This was very nice. She left to go home at eleven o'clock. We met the following evening and went for a long walk. We were very tired when we returned to the room. We both fell asleep, and it was almost daybreak when we awoke. She dressed in a hurry and left in a worried state of mind. There was nothing I could do to help. I hoped it wouldn't cause her any trouble. The following weekend I sent a message to her by one of the men. She wasn't at the dance, and I never got another chance to go on liberty before leaving New Zealand. I felt very bad about not being able to see her, and tell her goodbye.

The time we all dreaded finally came. All liberty was canceled; we prepared to move out. Everything in camp was hustle and bustle as all units prepared to depart the camp for good. The preparations continued night and day. There was a steady stream of convoys leaving the camp for Wellington, where they were loaded aboard ship.

Fox Battery departed without me; I was left behind on a working party to load the last trace of Fox Battery on trucks and to police the area. Once this was completed there was nothing to do, so I decided to go down to McKay's Crossing and see Bernnie before he left. His camp was already deserted; B-1-2 had departed. I didn't believe there were any guards on the main gate so I took a shortcut across the fence and went into the small town of Paekakariki dressed in dungarees. There were no other marines in town, so I wasn't worried about the shore patrol. I went to a small teahouse and had a bad cup of coffee and some cookies. Everyone in this town knew we were leaving. The people in the teahouse said that we were probably bound for the Gilbert Islands. I had never heard of the Gilbert Islands, so I didn't give that rumor much thought. When I got back to camp, I told the men what the people in town were saying. They said this wasn't news; Tokyo Rose had been saying the same thing on the radio. I was the only marine in Paekakariki that night, and probably the last marine to make liberty in that small town.

The following morning we loaded up and departed. The Paekakariki camp was no more; what was once a huge marine base looked like a ghost town. It was completely deserted. We passed through the empty guard house at the main gate and headed for the docks in Wellington.

I was very thankful to find we would be on a navy vessel, and not on another Dutch freighter. The food on the navy ship was nothing special, but at least we didn't starve.

When we shipped out of San Diego I stood on the afterdeck and watched the land fade away and wondered if I'd ever live to see the shores of the United States again. All kinds of thoughts flooded my mind. If I didn't come back, how hard would it be on the folks back home? If I did come back, would I be a cripple, unable to take care of myself? These questions were foremost in my mind. You can only imagine what fate has in store for you.

Now here I was again, on the afterdeck of another ship, watching the beautiful land of New Zealand fade into the distance. I've left behind someone I love very much, and will probably never see again. I don't have her address; anyway I would be afraid to write. It would only cause more heartaches for her and myself. This will be the past, a part of my life that I am forced to leave behind—a short span in two people's lives, but a closeness that I had never experienced before, and one I'll never forget.

The land faded into nothingness like a vapor trail; it was gone and could never be recaptured. The only thing that could be salvaged was the memory of a beautiful girl, the fragrance of new-mown hay, the odor of musky damp hair, and the sweet smell of the animal musk of two people embracing on a cool rainy evening. This image will remain with me forever.

Tarawa 1943

Life aboard ship is not a lazy man's life. You are always kept busy doing something—cleaning the heads, cleaning your quarters, or cleaning your own equipment. We had inspection of the quarters every morning at ten o'clock except Sunday. The inspection generally lasted until noon, and at that time no one was allowed in the compartment until the inspection was completed. If you failed the morning inspection, you spent the rest of the day correcting the deficiencies. You had a hard time staying topside. The sailors were always busy washing down the decks or performing some other maintenance task. I can still hear them yelling, "Hey marine, you can't stand there. Hey marine, you can't sit there; you'll have to move it."

After several days aboard ship, maps of the Gilbert Islands were painted on the deck of the ship. These would be used for our briefings for the upcoming campaign. We had a number of briefings. The Second Marines would make an amphibious assault on the island of Betio, one island in the Gilbert Islands chain. This island would later be remembered as Tarawa. In one briefing we were told that the First Battalion, Second Marines would be the assault unit landing on Red Beach One. This was Bernnie's battalion. Baker Company would be one of the assault companies. We were also told that this island was very small, and should be a pushover, being only a mile long at its longest point. I hoped for Bernnie's sake and for the men in the First Battalion that they were right. I didn't believe this because if Tokyo Rose and all the people in New Zealand knew where we were going it was quite certain the Japanese knew and were waiting. You don't send a large Naval fleet of ships and a division of reinforced marines to assault a mile-long island if it is going to be a pushover.

Blackouts were strictly observed on this ship. Mess and guard were doled out to everyone. As soon as you thought you could take a break, there was an abandon ship drill. We were busy doing something all the time. Looking back, it was probably a good thing.

Every morning you had to make roll call, and every man had to be physically accounted for. I was all for this; you never knew when a man might lose his footing and fall overboard. When we did get some idle time, it was spent reading or playing cards.

While en route to the Gilbert Islands, we made a brief stop in the New Hebrides Islands. On landing, I was fascinated by the huge mass of jungle growth. It came almost down to the shoreline. After penetrating the first wall of bushy growth, you'd come out in a clearing beneath some gigantic trees. We were on the jungle floor, and yet it was almost completely dark from all the foliage on top of the big trees hovering high overhead. This was quite a spectacle to me. I had never seen anything like it before, nor have I since. We only stayed for a short time, but it was good to get off the ship and just walk around on solid ground.

A heavy sea and an air bombardment were in progress when we reached the vicinity of the Gilberts. Fires and explosions could be seen all over the small island of Tarawa. Everyone was braced mentally for the morning assault.

Light from the heavy bombardment reflected off hundreds of boats and amphibian tractors rendezvousing at different points in the water, poised and ready for the first assault on the enemy. At the crack of dawn the first assault waves started for the shore. Our unit remained aboard ship, and we were notified that only one battery from our battalion was landing on Tarawa. Sgt. Prosser, being a good combat trooper, had volunteered Fox Battery for the assault landing, but for some reason we were not selected. Our forward observers would land on Tarawa, and we would land our guns on a small island south of the main island and deliver our fire support from there.

Later that evening the wounded started coming on board our troop transport. All of the hospital ships were full. There were so many wounded that troop transports were used as sick bays to handle the overflow of wounded. Our sick bay soon became overloaded, so the wounded were sent to our mess hall or galley. A section was roped off, and the mess tables were used as operating tables. I went below deck and stayed behind the ropes watching every kind of operation imaginable. All kinds of amputations were performed. I was an observer until they removed the testicles from one marine. This was enough for me. I went topside for some fresh air. One thing was certain in my mind. I could never be a surgeon. I don't know who performed these operations, or if they were really qualified to do them. All the ships overflowed with wounded marines, so I don't suppose there was any alternative. The operations on the tables were still in progress when we got the call to go to our debarkation stations. We descended the cargo nets into the waiting boats. As our boat was leaving, another load of wounded marines was pulling alongside the ship. It seemed there was no end to the casualties.

We landed without any opposition on a small island one mile south of the island of Tarawa. The code name for this island was Helen.

We went inland, dug the guns in, got them laid on the proper azimuth and ready to fire. Our forward observers were already on the island of Tarawa. We fired a few missions, though not as many as I had expected.

I still had not been assigned to a position in the gun section, so I was sent back to the bench on a working party, unloading ammo and supplies from the incoming boats. This meant working in the tropical heat with only one canteen of water every twenty-four hours. If you were in the gun section, one canteen of water would be sufficient, but under the hot tropical sun, it usually was a problem. I began practicing the same self-discipline we had been trained to do back in Camp Pendleton. Each morning we filled our canteens from old expeditionary cans of water. It had the color of rust and a very bad taste. We had no choice. We were forbidden to drink the island water because it contained parasites which could result in death or acute dysentery. Captain Johnson called this the GI trots. Some of the men were so thirsty they drank most of the water as soon as they filled their canteens. I was able to make one canteen last for the full time, and still have enough to give others a sip now and then.

One night while our working party was unloading the boats, someone yelled, "Condition red." This was being yelled up and down the beach. I had no idea what it meant. Everyone fled the beach, running inland. I found myself all alone. I put on my helmet, picked up my carbine, and started walking down the beach. I left the boats and the battle of Tarawa behind me as I continued south. I walked a long way down the beach and didn't see anyone. I turned and went inland through the underbrush for about thirty yards. I came to a small clearing and sat down with my back to the beach, and leaned back against a palm tree. I decided to stay there until I heard the sound of the men back at the boats. I sat looking across the small clearing. When my eyes became adjusted to the area, I could see two bodies lying just inside the clearing on the other side. You could make out the forms of the bodies by the moonlight. I assumed these were the bodies of Japanese that had been killed by a marine patrol. I continued sitting in the same position, not moving. Soon I noticed one of the bodies move, and get up; at about the same time the other one stood up. They were Japanese soldiers. I pushed my safety off and fired two shots at the soldier nearest to me, and snapped off two shots at the other one, as he bolted for the cover of the shadows. I jumped up and did a fast double-time through the underbrush back to the beach. In running back to the boats, I ran along the edge of the water where the sand was packed solid, where I could get some speed. I was holding the top of my helmet down with my left hand, and holding my carbine in my right. I made quick time getting back to the boats. I was sweating and completely out of breath. There was still no sign of anyone on the beach. I stood by the boats looking down the beach, not knowing if I had been chased or not. I learned something about myself, that as scared as I was, I could still think rationally. One thing I couldn't do was stop my hands from shaking. I looked at them, and tried to will them not to shake, but it was useless. It didn't matter what I tried; they continued to shake.

I believe the two Japanese soldiers heard me enter the area and had hit the deck. When I sat down in the shadows, and was quiet, they probably thought that I had left, and then had started to get up. I'm not sure what their mission was, but they sure got an unwanted surprise. I was still waiting, watching down the beach for any movement, when I heard someone yell, "Condition green," and everyone started coming back to the beach. I asked where they had been. They told me that condition red meant an air attack was imminent, and to take cover. When the warning is over, the all clear is given as condition green. This was news to me; no one had ever told me this.

When we finished unloading the boats, I went back to the gun section. I told Richard Zieske, one of the gun crew, about shooting the Japs. He said, "Big deal, go over there," pointing toward the island of Tarawa, "you can shoot all the Japs you want." I didn't mention this again to anyone.

When the tide was down you could walk from the island of Tarawa across the beach to our island. Some of the Japanese tried to escape the fighting by escaping across this stretch of beach.

The wind blowing across the island of Tarawa overpowered our island with the stench of the dead. There were so many dead bodies on the island that it was some time before they could be buried or disposed of. In the meantime, they were left to putrefy under a hot sun.

Late one night all hell seemed to break loose. The machine gunner on the beach was firing as if he was being attacked. Everyone ran to the beach while the gunner fired across the beach connecting the two islands. The CO arrived out of breath and asked why the man was shooting. The gunner said, "The Japanese are running around on the beach in their white skivvy drawers." The colonel looked across the beach with his binoculars, turned to the gunner, and said, "There is nothing out there but a bunch of seagulls flying around about waist high. Now everyone get back to your duties." This sentry wasn't the only one to see mirages while standing a late watch in a combat zone. Your eyes can play tricks on you. One night I watched a stump climb a tree. The only way I could tell the difference was not to look directly at the object, but look away to the side and observe the object out of the corner of the eye. When you are scared and alone, the shadows can play many kinds of tricks on you.

A small barbed wire stockade was built a short distance from our gun positions to hold the Japanese prisoners captured on this island. I passed the stockade one day and was surprised to see a marine prisoner in the stockade with the Japanese prisoners. The sentry told me that when the marine was wading out to a boat, he was behind his CO when he pulled his .45-caliber pistol out of its holster. The CO claimed the marine was going to kill him, but the Marine insisted he removed the pistol to keep it from getting wet. I never heard what happened to this marine, as he was not in our unit.

The island of Tarawa had been secured for five days when I was sent back to the beach for something. I decided to walk down the beach and try to locate the small clearing where I had shot the two Japs. I hoped to get the rifles or bayonets for souvenirs. I found the clearing and one dead body. His body had swollen to twice its normal size. His rifle and bayonet were both missing. I looked beyond this body for the other Jap soldier but I couldn't find him. Possibly I missed him because I was firing as he retreated to the darkness of the underbrush behind him. I cut across the island on my way back to the battery. I had only walked a short distance when I came across two more dead Japanese soldiers. From the looks of the bodies, they appeared to have been slain at about the same time I fired at the two soldiers in the clearing. I don't know who killed these two, but their weapons were also missing.

An important milestone took place in my life after the battle of Tarawa. I shaved for the first time in my life, scraping the fuzz off using the tin mirror, razor, and tube of Barbasol shaving cream passed out by the Red Cross. It was about time, after all I was now a veteran of one campaign.

I had been worried about Bernnie since D-day. I worried that he might be one of the wounded that had been brought aboard our ship. I located the graves' registration unit and was told that there had been so many casualties that it would be weeks, and maybe months, before they would have all the names of the killed and wounded recorded.

When the Second Marine Division left the island of Tarawa, it was said that you could take a rock and throw it from one end of the division to the other. This was all that was left of over twenty thousand marines who had made the heroic assault on the heavily fortified island of Tarawa.

Out of over four thousand imperial Japanese marines who had defended the island, only a small number were captured. The rest were either killed or had committed suicide.

After nine days on the island of Helen, we embarked aboard the *USS*

Harris. It took us about a week to reach the port of Hilo, Hawaii. I dreamed of palm trees, nice beaches, and beautiful hula girls. I could just picture myself in a camp a short distance from Hilo, and being able to walk into town on liberty. I thought this would be paradise compared to other marine camps.

We debarked and got the guns and trucks lined up in convoy formation. When all was ready we started for our camp area. The convoy got to the edge of town and just kept on going and going. We went along a winding road that had been cut through huge lava rocks. It seemed we were moving through a mountain of frozen molasses. We finally stopped about ninety miles from Hilo at a place called Parker's Ranch. We were right in the middle of a semi-desert area, with fine sand blowing across the convoy. What a disappointment! It was mostly flat desert with a lot of cactus. If this was to be our training camp we would have to build it. There was nothing here. As soon as we unloaded we set about building the camp, putting up tents, hauling coral rocks for the tent decks, digging out houses, and constructing a water system.

The sand blew all the time; you couldn't escape it. Now and then we would have a big sandstorm.

We used a supply tent for the mess when we were served our 1943 Christmas dinner. As you went through the chow line the flaps of the tent blew, kicking up sand and dust so thick you could hardly see. The food was very gritty. I remember scraping the gravy off the mashed potatoes to eliminate some of the gritty sand. It was a Christmas feast I'll never forget.

Our camp was called Camp Tarawa. This would be our home and training grounds for the next six months.

Our pay records were still in New Zealand, and no one knew how long it would take to get them to Hawaii. Since no one had any money to go on liberty, everyone tried a little harder to make the camp more liveable. What time we weren't working on the camp, we stayed busy cleaning the guns or having gun drills. We didn't know what our next campaign would be, but we got prepared as best we could. The only recreation was a hike into the small town of Kameula and a visit to the USO, or an occasional payday poker game.

It was over a month before I had a chance to locate the Second Marine camp. I still had heard nothing from home or from Bernnie. After a long and very hot walk, I located B-1-2. I entered the first sergeant's tent and came face-to-face with Jones. He was one of the men in my brother's machine gun section. By the process of elimination, Jones and Taylor were the two oldest men in the company. Jones was acting first sergeant, and Taylor was the acting gunnery sergeant. They only had about sixty men who had survived the battle; the rest had been killed or wounded. Bernnie had been manning one of the .50-caliber machine guns on the amphibian tractor when it was hit by enemy mortars. He was in the first wave, and was lucky to have been in an amphibian tractor. The men in the boats got hung up on the reef, and had to wade ashore under heavy enemy fire. Jones told me Bernnie had been hit in the face; he didn't know how bad. Once the tractor was hit, Jones and Taylor took the machine gun inland, got it set up, and started firing. Taylor had been wounded, and Jones had received the Silver Star for this action. They never heard what happened to Bernnie, and didn't know when he was evacuated from the beach.

The battle for Tarawa had been over for about two months when I finally received a letter from home. Bernnie had been in the hospital in Pearl Harbor for quite some time before being sent to the hospital in Oakland, California. He had received over twenty operations trying to restore his face. He lost the sight in one eye. Later he was placed on limited duty, and finally discharged from the Marine Corps.

When I did get a letter from Bernnie, I learned he had remained wounded on the beach for the entire day before anyone could get him off the island for medical attention. Every part of the island was subjected to enemy fire, making some evacuations almost impossible.

I was still dreaming of pretty hula girls when our pay records finally caught up with us. Most of the men made liberties in the small towns of Kamuela and Honokaa. A few men made liberty in the larger town of Hilo, but all who did came back disappointed, so I never went into Hilo on liberty. I was disappointed with my liberty in Honokaa one day, so I went to the next small town to look around. This was the town of Paauilo, much smaller than Honokaa. I met a pretty Japanese girl. She said she was eighteen, but I had my doubts. She told me her folks would never let her date a service man. She asked me to meet her the following Sunday evening behind the church. This proved to be a choice meeting place. It was a large church, and the back was very secluded, with large hedges surrounding the rear. Our tryst lasted for a number of Sunday evenings. I became very attached to her. One Sunday she didn't show up. I waited until late in the evening before giving up and catching a ride back to camp. I was there waiting for the next two Sundays before finally giving up. Something happened; possibly her father found out. To this day I hope she didn't end up in any trouble because of our dates. She was a doll, and I considered myself lucky to have a sweetheart while in Hawaii. Girls were scarce, and I didn't know another marine who had a girlfriend on the island. It was a short romance, but it was sweet and one I'll not soon forget.

We made a number of maneuvers out in the boondocks of Hawaii. Boondocks are what the marines call lonely stretches of beaches, swamps, or any other God-forsaken place, like the outback of Australia. There were similar places on this island, and we went into them for artillery practice. It seemed every time we fired a white phosphorous, or smoke round, it would ignite the dry grass and brush. All we ended up doing was firing one round, and spending the rest of the day fighting prairie fires. One of the fires got out of control, and the whole Second Marine Division spent the best part of three days trying to put it out. I did my part on the side of a small hill, using burlap bags and buckets of water for over thirty hours without a break. Just about everyone in the division fought the fire around the clock.

We made numerous hikes out into the sandy boondocks, and always a cloud of dust hung over the whole battery while on the march. I felt sorry for the feather merchants, the smaller men on the tail end of the column. These men ate dust for the entire march. Riding in the back of a 6x6 truck gave no relief from the infernal volcanic dust, as the tires from the truck would continue to churn the dust up, and into, the back of the trucks.

I was put on many working parties from the very first time we dismounted on this desert ranch. I became very proficient at constructing four-hole outhouses. After digging the holes, some of the other men and I would construct the wooden structure over the holes. I lost track of how many I built or helped to build.

Once while in this camp we had a large infestation of crabs; it seemed everyone had them. The corpsman checked each man with a flashlight before he went on liberty. If he detected any sign of lice, you were restricted until you were treated, checked, and cleared. So as not to miss any liberties, I shaved off every hair on my body except on my head. After this, I never had any more trouble with these parasites. All of the men who were restricted to the base spent their time washing and swabbing the seats of the head, or toilet, with disinfectant, to try and eliminate the spread of these insects.

Parker's Ranch threw a monster of a barbecue for all the marines of the Second Marine Division. There was plenty of food and a rodeo. This was a lot of fun with a carnival atmosphere. Parker's Ranch, the host for the marines' party, was most gracious and considerate; even so I was still glad to hear that we would be leaving the island very shortly.

This island left me with some fond memories. In a way I was glad to go, and in another way, it seemed I belonged to this island.

I saw more desert and cactus in my short stay in Hawaii than I had ever seen in Texas. I must say this was a far cry from the palm trees and hula girls that I had envisioned on arriving in the islands.

At last we boarded trucks in a convoy that would ferry us back through the dusty roads, along the edge of the molten volcanic rock, down to the port of Hilo. Most of the day was spent getting the guns and equipment loaded aboard an LST to make another rendezvous with an unknown destiny.

Mariana Islands

After a brief stopover in Honolulu, we joined a naval armada that stretched from one horizon to the other. Every kind of naval vessel was represented: battleships, cruisers, flattops, troop ships, LSTs, and destroyers. We had not received any word on where we were bound, but from the looks of this ocean dotted with ships, it was quite evident to everyone that this would be a massive invasion.

Our LST was small and very crowded. It was hard to find anything to read. I spent many hours riding the bow up and down, watching the flying fish as they skimmed along in the front of the ship. At times the bow of the ship made a deep dive and then salt water sprayed across the deck. I enjoyed this, and thought it was fun, but many of the men thought I was crazy. A lot of the men got seasick and spent most of their time close to the side of the ship, gagging when they had nothing to come up. I felt sorry for them, as most of them would be miserable for the entire cruise. I had many months aboard different ships and boats, some in stormy or choppy waters, and was never seasick. I knew only a few men like myself who seemed to be immune to seasickness. I could never understand how or why we were any different from the other men. The wake of the mighty navy armada continued its path through the Pacific. We passed the vicinity of Tarawa, and it triggered memories of all the men who had fought and died there. There were many heroes honored in this battle, like Lt. Hawkins. For every man who was honored, there are many more heroes whose names will never be known or remembered. The deadly past lays buried in the faded wake of the ships. It was gone forever. The future, whatever it was going to be, still loomed just ahead of this gigantic convoy.

Many of the men and I, being worried about the next campaign, could not sleep. Because it was always hot below, I spent a lot of time on the top deck. Sometimes I would take a blanket, go topside, and find a cool place to sleep. We were very lucky to be aboard an LST that was being manned by the Coast Guard. The crew was really good to the marines. They provided coffee all night for the men who could not sleep.

Blackouts were strictly observed by everyone because we were always worried about torpedoes striking our LST.

Soon we started our briefing for the next amphibious landing. It had been a secret up until now; the mystery was gone. Our objective was to be the island of Saipan in the Marianas. According to the briefing, the Second Marine Division would land on the west side of the island. The Fourth Marine Division would land on the east side, and the Twenty-seventh Army Division would land south of the Second Marine Division, move inland, turn north, and help consolidate the front lines between the two marine divisions. This was the basic plan. We were more concerned with the landing of our own unit, and what our mission would be. According to the briefing, this island would be well fortified and defended. Our mission was to deliver artillery support for the Eighth Marines with the 75-mm pack howitzers.

This island would be shelled and bombarded prior to our landing, and I hoped it would have a more devastating effect than it did on the island of Tarawa.

We arrived in the area late one evening; the planned assault was to be the following morning.

Our LST moved along at a very fast clip; general quarters had been sounded, and everyone was at their battle stations. I manned a double .50-caliber machine gun on the bow of the ship. We expected to come under air attack at any second. I had a navy commander manning the phones right beside me, telling everyone not to fire until he gave the word. A Japanese dive bomber came diving in very close to our ship. I tracked it down with my twin .50, waiting for the order to fire, but none came. It was close enough to hit if the order had been given to fire. The bomber came down and dropped a bomb that exploded just in front of another LST moving parallel to our port side. The bomber pulled up, then moved away from our ship and out of range of my machine guns. The order was then given to commence firing. It seemed every gun in the fleet commenced firing at the same time. The LST that had been missed by the bomber received one round from one of our ships, which ignited the fuel stored on the top deck. There was a tremendous explosion on top of the ship. It went dead in the water as secondary explosions sent large plumes of smoke and debris high into the air. I have no idea how many marines were lost on this ship, but it had to be a huge number because it was loaded to capacity with troops, just like our ship. The Jap bomber accomplished his mission with the hep of some trigger-happy gunner on one of our ships. A short time later, well after dark, we stopped. We were in the staging area for the initial assault the following morning.

A massive sea and air bombardment continued all through the night, softening up the beach to try and give the marines an edge in establishing a beachhead.

Reveille sounded at two o'clock in the morning. It wasn't easy to eat breakfast that early. Everyone was too worried about the landing to enjoy the meal. I sat and drank coffee as long as I could. The paging came over the PA system calling our unit to the debarkation station. We debarked down cargo nets into the boats.

At the last minute I was informed that I wouldn't be going in with the gun section. I was put on a working party ferrying in ammunition and supplies. This was a big disappointment; I had hoped to land with the gun section. I stood on the rail watching the boats circling in the water in their respective rendezvous points. Our forward observers would go in with the infantry; the guns would land in a later wave. It seemed our boats were in the water a long time before they started for the beach. I believe the guns of F-2-10 were in the ninth wave to go ashore. Just as I was about to depart the ship with the shore working party, we were informed that casualties on the beach were already heavy and that they continued to mount. We started for shore at the break of day. We received sporadic artillery fire on our way to the beach. The fire was haphazard and didn't appear to be aimed at any particular target. I was scared, not knowing what to expect once we hit the beach.

I have never written poetry, but the somber breaking of the day on such a dire occasion inspired me to jot down the few words that came to me. I kept this small poem for years and will recite it later in this book.

The beach was in shambles with disabled equipment and dead bodies of marines and Japanese scattered all over the place. Wounded were being brought to the beach to be evacuated to the hospital ships.

The beaches were still under enemy artillery and small arms fire, but not writhing fire as experienced by the previous wave of troops on landing. While we unloaded the ammunition, others carried and stacked the ammunition in a palm grove near the beach. All kinds of ammunition, demolitions, and explosives were stashed in this grove. This would be the ammunition supply dump for the Second Marine Division. Our first night on the island was scary. We continued working although we were all very tired, but no one went to sleep. Most of the time I worked I could hear our guns firing in the near distance. They were the only artillery on the island, and had their hands full delivering both fire support to the infantry and firing offensive at the big guns of the Japanese that were located in caves that had been dug back into the mountain.

We continued working all of the next day. Late that evening we were told to report to our respective units. It wasn't hard to locate our guns as they were only a short distance from the ammo dump.

On reporting to the gun section, I learned of some of the casualties of the Second Battalion, Tenth Marines: Col. George A. E. Shell, the battalion commander, had suffered wounds, as had Capt. Johnston, our battery commander. Our gunnery sergeant had been wounded, and Pfc. Mannon, our switchboard operator, had been killed. There had been others wounded. We only had two officers left—Lt. Pape in the fire direction center, and Lt. Meyers, acting battery commander.

I had a hectic first night back at the gun section. I would spell some of the men so they could get some rest, but no one could sleep. We were still getting incoming artillery fire from the enemy. Whenever you heard the screaming sounds of incoming artillery, you scrambled for your foxhole. It landed all around our gun pit; some of our men had some very close calls. Two of our men remained in the gun pit during one barrage. When it was over they discovered their foxhole had had a direct hit. There was a gaping hole full of sand. Then they retrieved their packs from the sand everything was full of holes.

When we fired our guns, the enemy fired at us, and everyone scurried for cover. This was repeated over and over. Finally the 105-mm battalion landed, and between their battalion and ours, an artillery duel took place. When the other unit was under fire, our guns would start firing. When we came under fire the other artillery battalion jumped into action. This duel went on for some time. I was still busy making trips back to the ammo dump for more cloverleaves of 75-mm ammo. A cloverleaf was three rounds encased in a wooden container. It weighed about seventy-two pounds. I was glad when the enemy artillery slacked off. It made it easier to get to the dump and back.

On the third night the 105-mm battalion pounded the Japanese positions. This gave us a short break. Most of the gun crew had not slept in three days. I was dead on my feet, but the smell of the dead was enough to keep you alert and awake. We had a log behind the gun to take the brunt of the gun's recoil. I lay down with my head on this log and went to sleep. I must have dozed for over an hour when I was awakened by a loud explosion that threw me into the air. I came down and struck my head and shoulders on the log. I could taste blood and, being dazed, thought that we had received an air burst from the Japanese artillery. I stood up and looked around. There was a fire, an exploding small arms ammo in the palm grove. The ammunition dump had exploded thirty-five yards from our gun pit. I heard a lieutenant yelling for men to get over there with shovels.

As I watched, men ran with shovels to put out the fire. They shoveled dirt on the exploding ammunition that sounded like firecrackers. Cpl.

Kendon, the assistant section chief, told me to get a shovel and give them a hand. I said, "I never heard of anyone fighting an ammunition fire with shovels." Kendon said, "Neither have I. It sounds dumb, but get over there and do what you can."

The explosion had shaken everything up, but I finally located a shovel. I walked to the top of the gun pit and started for the fire. I saw a bunch of men throwing dirt on the fire. All at once a blinding flash followed by an explosion hit me with such force that it blew me up above the gun. I came down on the back of the gun, striking my back on the trail and my head on the trail log. I stood up and looked toward the ammo dump where the small arms ammo was still exploding, but the men were gone. All of the marines had been blown up; many of the bodies were strewn all around our battery area. Many were dead or unconscious; a few were still alive but bleeding from their nose, ears, and mouth due to the concussion of the blast. Some of these died after we got them into the gun pit. We had to work fast, as the ammunition dump kept exploding, scattering heavy artillery projectiles all over the area.

I was still bleeding from the mouth and ears as I was sent out in front of our position to guard against a possible banzai attack. Every time I saw a flash I ducked down in my foxhole only to have it leveled by the blast. Each time this happened I would find myself on top of the ground, and hurriedly set about digging it out again. Each explosion leveled it again, and heavy projectiles dropped all around me. I was afraid one would land on top of me. This went on all night.

When daylight came the ammo fire had died down, and the large explosions seemed to have run their course. I stood up but had to be extra careful where I stepped because the ground all around my foxhole was covered with unexploded artillery and mortar rounds. Most of the men who had been fighting the fire had been killed, including the lieutenant. The final count of the dead was thirty-five. A Jap sniper firing into the ammunition dump had exploded two thousand pounds of gelatin dynamite. From then on it was a chain reaction until the last ammo was destroyed.

After losing our ammo dump, we got our ammunition from the Twenty-seventh Army Division's dump. This dump was about 500 yards down the beach from our position. This involved dancing the boondocker ballet, that is, running a gauntlet of enemy artillery fire going to and then returning to the gun section with one cloverleaf of 75-mm ammo. I would leave our section and run the 500 yards with an occasional dive for cover if I heard an artillery shell screaming in. I would grab a cloverleaf of ammo and then double-time back to the battery. If I heard incoming artillery, I dropped the ammo and dove into the nearest hole or bunker. Some of these holes were already occupied with either dead marines or Japs. If you needed the protection, it didn't seem to matter who you shared the hole with. Once the enemy artillery round exploded, you were up and moving again. The tropical heat had decomposed the bodies, and the smell was sickening. If you landed in a hole with one of these bodies, you had no choice but to stay until the enemy artillery abated. After a number of trips, you began to be more selective about which holes you dove into.

At one time we got the alarm that a platoon of tanks had broken through the lines and was coming down the small railroad tracks on our left flank. We got the guns out of the gun pits and set them up facing the oncoming tanks. We prepared for direct fire, but they never reached us. Two men with bazookas and a .50-caliber machine gun section stopped them dead in their tracks. These were light tanks, and the .50-caliber machine gun cut the turret completely off of one tank.

The big guns on top of Mount Tapotchau had been silenced. One of our scout sergeants, Sgt. Dees, and his radio man were instrumental in quieting these big guns. These guns had been built to roll out on rails, fire the gun, and then roll them back into the caves on the side of the mountain. Once the guns were withdrawn into the cave, large coconut log doors would close behind them, thereby giving Japanese gun crews protection while they reloaded.

I still had not been assigned to a permanent position in the gun section. It was at about this time that I was sent to the front lines to assist Cpl. W. D. Bell in his liaison duties, maintaining communications between the artillery units and the forward observers on the front lines. The first night on the front line with the Eighth Marines, Bell and I decided to dig a foxhole together. While we dug, I told him he smelled worse than a dead goat. He stopped digging, looked at me, and said, "I was just about to tell you the same thing." We ended up digging separate foxholes.

We set up on a small hill, just south of the town of Garapan, the biggest town on the island. We were observing our planes strafing the outskirts of the town when one of the planes received a hit. It was smoking, losing altitude, and coming straight toward our hill. The pilot bailed out at about 200 feet above the ground; his parachute only partially opened. We expected the plane to crash on our hill, and we all made a dash to the reverse slope for protection. The plane crashed in flames just in front of our position. A patrol was sent out immediately to retrieve the body of the pilot before the Japs found him. We found him still alive but badly hurt. We removed him on a stretcher and sent him to the rear for medical attention. I never heard if he survived his injuries.

Soon after this we moved up to the edge of the town of Garapan. It was here we were ordered to wait for the Twenty-seventh Army Division to even up the lines. The army had some rough terrain to cover and had fallen behind. This left our right flank exposed to some extent. Patrols were sent out routinely to keep a check on our right flank. While on one of these patrols, Bell and I spotted a two-story frame building. It was located quite a ways to the north of Garapan—just about as far as you could see with the naked eye. The building was located next to a large grove of trees. Wires led into the top corner of the building from every direction. We thought that this might be an installation of the local telephone company, and that they had to have a switchboard to accommodate so many lines. A switchboard would have made our job a lot easier. We noted the terrain with a drainage ditch winding up through the valley. It was in a north-south direction and passed a short distance from the frame building. We decided to go after dark and try to get the switchboard.

That night around nine o'clock we got the password from the sentry nearest the drainage ditch and asked him to watch for our return. I had estimated the distance to be about one and a half to two miles. Bell and I were both in top physical condition and the distance could have been farther. Enemy mortars exploded off to our right, and we heard machine guns firing in the distance as we departed. Every time a flare went off we froze in our tracks. Those were the longest stops we made on our journey. Once the flares were extinguished, we would take off again. I had thoughts of going back, but we were already about halfway, so I kept on going. I wasn't wearing a helmet or a cartridge belt. I had my carbine, with one-ten rounds magazine in the carbine, and another one in my pocket. I had my K-bar knife hanging on my web belt. Bell wasn't wearing his helmet either, but he was wearing a pistol belt with lineman's pliers and a .45-caliber pistol on this belt. We came out of the drainage ditch about fifty yards east of the building. It wasn't a house; it looked more like a barn with double doors. Aside from the closed double doors on the front facing the marine lines, there were no other openings.

We cautiously approached the northeast corner of the building. The double doors to the rear were open. We could hear voices in the dark interior. Peeking around the corner, we saw three or four soldiers smoking. They were talking to soldiers on the other side of the room. We could not observe them, but there had to be about a squad of men in the room. On the opposite side of the doors from us was an outside wooden stairway leading up to the second-story room. Since we could not cross the opening of the double doors without being seen, we backed off and went about seventy-five yards north, made a U-turn, and came back to the foot of the outside stairway. At the top was a single door which was closed. We didn't know what to expect as we eased the door open. I had my K-bar knife ready but there was no one in the room. Candles were burning and the smell of burning punk used to ward off mosquitoes punctuated the air. There was a beautiful switchboard wired up and in operation. It looked brand new and had handles on both ends, like a metal box of .50-caliber machine gun ammo. I carefully disconnected wires while Bell cut them with pliers. We picked it up and started for the door when Bell picked up a hand-cranking generator and laid this on top of the switchboard. I had to shoulder my carbine and back down the stairs. I held the switchboard with one hand and tried to keep the generator from falling off with the other. Bell was a short man, and I was six feet tall, so we were able to keep the switchboard level as we descended the stairs. This was very unnerving, as any Jap coming out of the room below would naturally be looking right at my back. If that had happened, we wouldn't have stood a chance.

At the bottom of the stairs, we went north toward the Japanese lines for about seventy-five yards, turned east, went directly to the drainage ditch, and retraced our steps back to our front lines without any mishap. We arrived back at our lines sometime after midnight, because there was a different sentry on post. I did not sleep that night. I was only eighteen years old at the time, and I just knew the Japanese would pull a banzai attack against us once they discovered the switchboard was missing.

Bell and I spent two days trying to figure out how to adapt it to our use. We were handicapped because wording and instructions on the switchboard were in Japanese. It was real shiny and the prettiest piece of equipment I had ever seen. We were still studying this piece of equipment when an Eighth Marine lieutenant came by and told us we would have to turn it in to be checked by the censors. The lieutenant put it in the back of his jeep, and we never saw it again. About two days later I returned to F-2-10. I don't know what happened to Bell, or when he returned to the States. I never worked with him again.

I went on full retirement at the age of fifty-five in 1978. After my retirement, I began reading in my leisure time. I bought a book by Lempke titled *Marines in the Pacific*. In reading his account of the battle of Saipan, he stated, "The Japanese commander, Lt. Gen. Yoshitsugo Saito, was preparing a massive counterattack against the marines, driving them into the ocean. At this time Commander Saito had the manpower to do so, but about halfway through the battle he lost all communication with his forces." I have no idea what was researched, or where Mr. Lempke got his information, but I can assure him, or anyone, that this information is correct.

It would be great to say we stole the switchboard to sabotage the Japanese communication system, which we did, or that our mission was to save American lives, which we most certainly did—my life included. However, this would not be the truth. Stealing the switchboard was a devastating blow to the Japanese coming as it did at a most critical time for the Japanese commander. I regret to say that Cpl. W. D. Bell and myself did not undertake this mission as a sense of duty, but to secure a switchboard for our own use, thereby making our work a little easier. This was our sole purpose and motive at the time.

This information on stealing the switchboard was sent to the Marine Corps Headquarters. I received the following letter from Danny J. Crawford, Head Reference Section, History and Museums Division, Headquarters, United States Marine Corps.

"This is in response to your letter concerning your participation in the battle of Saipan in World War II. We have placed your detailed and informative letter relating the exploits of Corporal Bell and yourself during the battle for Saipan in our World War II reference files covering the campaign. I am sure future researchers working on the battle for Saipan will benefit from your letter. Thank you for taking the time to share this information with us. Your interest in the Marine Corps History is appreciated." Signed Danny J. Crawford.

The island of Saipan was secured in thirty-one days; it was seventeen days after we had taken the Japanese army's central switchboard, severing all communications between the Japanese commander and his army units.

I have often wondered whatever happened to the Japanese soldiers who were supposed to be manning the switchboard. I'm sure they caught hell for taking a short break from the war. Marines weren't making night patrols so the Japanese took advantage of this to get some rest. Bell and I unexpectedly caught them off guard.

Things had quieted down some when I arrived back at the gun section. Every so often the enemy fired a few artillery rounds into our area, but most of these rounds were not fired in barrages. They were mostly fired to harass us, and did little damage.

We advanced our howitzers to a different position, closer to the town of Garapan. Our machine gun sergeant had to blast trees down in front of our positions to give us a field of fire. Once the trees were down and the guns dug in, we continued to deliver fire support to the Eighth Marines.

Eating anything on this island was a two-handed chore, that is, eating with one hand, and fanning the flies away with the other hand. Sometimes the fly stuck to your food before you got it to your mouth. The flies and mosquitoes were busy day and night. No matter what you did, they would get to you, especially at night while on watch.

Every night while we were in this position, a Japanese twin engine Betty Bomber flew over and dropped bombs on island-targets. The anti-aircraft guns saturated the sky all around the plane with bursting shells. Terrific fireworks, but the plane, as slow-moving as it appeared, would complete the mission and leave. This was repeated every night for over a week. The lone pilot probably lived to a ripe old age.

We had been in this position for over a week when we learned the

Second Marine Division had been squeezed out by the Twenty-seventh Army Division, due to the taper or contour of the island. We were to remain in our position and give fire support to the Twenty-seventh Army Division if they needed or requested our support.

It was late in the evening when the weary, dirty, and battle-scarred men of the Second Marine Infantry Division came marching down the dusty coral road and on south past our gun positions. They marched twenty miles to the south of the island for a much-needed rest. Their combat prowess against the Japanese was unquestionable. Everyone in Fox Battery was glad to see them get a break from the front lines.

A short time after they passed our position, and just before dark, the Third Battalion, Tenth Marines, a 105-mm howitzer battalion, moved up to about 700 yards in front of our gun emplacements. Once they were dug in, and had their guns ready to fire, we would leapfrog our smaller guns up to a position somewhere in front of the larger guns. This was standard procedure, but on this evening it was very late when the Third Battalion completed getting into position and getting the guns dug in. We were told to have everything in readiness for our move early the following morning.

Along about midnight we were alerted that the Japanese had broken through the Twenty-seventh Army Division, and the army was in full retreat. Aircraft received word to drop rubber life rafts to soldiers who were trying to swim to ships anchored offshore. The only thing that stood between our battery and the Japanese army was the Third Battalion, Tenth Marine Artillery unit, who had just made the move up ahead of us. We immediately set about to intercept a massive Japanese banzai or counterattack. Each gun section broke out their canister rounds of artillery ammunition, made to be fired directly at attacking troops. Each gun section was given a case of grenades. It was very tense, sweating, sitting, and waiting. I had my carbine ready with two-ten-rounds magazines, and one extra box of ammo. I was wishing I had an M1 rifle with a bayonet. It was too late now to try and get anything; I would have to fight with what I had been issued. The Third Battalion, Tenth Marines up ahead of us had no choice but to brace themselves and be prepared for the onslaught of the Japanese army.

Around about midnight shooting started up in front of our position. It sounded like all hell had broken loose against the Third Battalion, Tenth Marines. Guns, grenades, small arms, mortars, and machine guns all fired simultaneously. We waited. If the Japanese broke through the 3-10 positions, F-2-10 would have a war to fight. We sat helplessly, waiting and listening to the raging battle up front. The noise of the battle would fade, only to reach a new and higher crescendo, then decrease, only to rise again. About daylight the noise of the battle started dying down. The fight was ending, but we still didn't know who had won. The F-2-10 was still poised to intercept the Japanese army once they overran the Third Battalion, Tenth Marines' position. We were still waiting when we heard the Third Battalion, Tenth Marines had stopped the Japanese banzai attack. Not one Japanese reached our position. The Third Battalion, Tenth Marines, an artillery unit numbering about 500 men, fought and beat over 5,000 Japanese soldiers in a desperate banzai attack. Each of their gun pits ended up being a separate fort. Some of the cannoneers stood in the middle of their gun pits and, using their rifle butts, batted the Jap grenades back to the Japs. The Third Battalion paid a heavy price for its bravery. Out of approximately 500 marines, only about 160 men survived that assault. They had fought and had held their position well.

During and after the battle, the search was still on for the remnants of the Twenty-seventh Army Division, who had deserted the front lines.

The Third Battalion, Tenth Marines received the Presidential Unit Citation for this battle, and details of the battle itself were later covered in *Leatherneck*, a magazine.

What a surprise when early the next morning the Second Marine Infantry Division passed our position to go back north again. They would take up the positions vacated by the Twenty-seventh army. They didn't get much of a rest after having marched all the way south the evening before.

At about this time, our guns slowed down on their fire missions, so I decided to go up to the front lines and visit my brother's old buddies, Sgt. Jones and Sgt. Taylor in B-1-2. As I walked north along the dusty road, trucks loaded with dead marines from the battle of 3-10 were going south. The bodies in back of the trucks were stacked to the top, with only the boondockers of their feet visible. After one ride and a long dusty hike, I finally found B-1-2. Sgt. Jones and Sgt. Taylor were in good shape, but they complained about the army and its retreat. I told them we had heard the army had run out of ammunition and was forced to retreat. They took me to their foxholes and showed me that just about every foxhole had a machine gun, and all the necessary ammo. These arms were picked up from where the army had abandoned them in the mad dash to the rear.

The position of B-1-2 was about two miles north of where Bell and I had procured the switchboard over two weeks before.

I had to cut my visit short as Jones and Taylor were organizing a burial working party for the dead bodies in the area. When I left they were still griping. They didn't mind killing the enemy, but they didn't relish the job of removing and burying the dead the following morning.

On the long walk back only a few vehicles passed, but they didn't offer me a ride. I marched past the Tanapag Harbor locale, kicking up dust hiking south. Six Japanese soldiers, captured by a patrol of marines during mop-up operations behind the lines, were being marched to a prison stockade. Shortly after, three marines came out of the shrubbery and onto the road. The three marines were carrying a pretty young Japanese girl dressed in a very clean kimono, which contrasted sharply to the filthy dirty dungarees the marines were wearing. She was unconscious. The marines asked me for directions to the nearest aid station. I pointed them toward Tanapag Harbor, which was the nearest one in that area. The three marines carried her down the road as fast as they could go. One marine supported her head, the other her back, and the third her feet and legs. They were being very careful, carrying her like a fragile doll. I watched until they were out of sight, then continued my trek south and back to Fox Battery. The battery was still in the same position. The fire missions were very few; it would be an infantry operation from this point on. A short time after my return to the battery, the island of Saipan was declared secured.

When I became very sick, I thought perhaps I had caught the flu. I ran a high fever, and every joint in my body was stiff and ached something awful. The battalion doctor was notified, and he was to meet me at the battalion headquarters. I faintly remember him telling me to take a couple of APC pills and get some rest. While he was talking to me, I passed out. When I came to, I was riding in a jeep with the driver trying to find the field hospital. After what seemed to be an eternity he located the field hospital. It was simply two tents connected end to end. I vaguely remember checking in.

When I came to, I was lying on a hard canvas cot with no mattress. All I had was my blanket and poncho. I ran the gamut of fever and chills. My mind faded in and out. I remember someone pouring some grapefruit juice in my canteen cup. My body was extremely sensitive, and the hard canvas cot felt like coarse sandpaper when I moved or turned over. On the third day, I was starving and asked for something to eat. I was told the only way to get something to eat was to take your mess gear, go outside, and get in the chow line. That evening I went out and got in a long chow line. I remember the sun was unbearably hot. I was about halfway through the line when the ground came up and hit me in the face. When I woke up I was back on the hard cot. I did get some more juice.

The following morning I went through the chow line and finally got something to eat. At ten o'clock the doctor came in and said I had contacted dengue fever, a kind of virus carried by mosquitoes. I was still very weak and had a hard time standing. My joints were still stiff. I heard the doctor say he was releasing me to go back to my unit. The hospital had no transportation, and I had no way of notifying my battery. I shouldered my pack and left the hospital.

I was tired and completely worn out when I reached the old gun positions. Fox Battery had departed; there were no marine units around. I drank all of the water in my canteen and sat down to rest. After a short break, I went south and, after a long hike and many questions, I successfully located Fox Battery. Most had boarded ships or boats to make the assault landing on the island of Tinian. D-day was to be the following morning. I found one Landing Craft Vehicle Personnel (LCVP) that had a working party for our battery on board, so I joined them. I spent the whole night topside, with a strong cold wind sweeping the deck. I thought I would freeze to death. All I had for protection was my blanket and poncho which was far from adequate for such inclement weather. The following morning I went ashore with the working party to unload ammunition for our battalion. The Second Marine Division had already secured the beachhead. I had no way of getting word to Fox Battery, so I stayed with the working party. Pits had been bulldozed out for ammunition storage. The trucks were loaded on the beach and brought to the ammo pit. It was my job to carry ammo from the trucks into the ammo pit. We didn't get a break; there were always trucks waiting to be unloaded. My joints were very stiff when I started, but after twenty-four hours on this working party I began to limber up. When I got to the gun section I was worn out and was hoping for some rest, but I immediately pulled watch to relieve one of the gun crew. There was no reason to complain or argue. No one had it easy. I was no exception. When I was told to do something, I did it.

It was almost impossible to dig gun pits and foxholes in this hard coral rock, but we did the best we could. We spent most of the day chopping down the sugar cane in front of our positions. The cane had been burned off, and all the time that you chopped the stalks off, the ashes and the burned remnants of the leaves would go down your collar. This, combined with the sweat from your body, was pure torture. Everyone bitched and complained about all the time spent cutting the cane. The only reasons given for doing so was to enhance our field of fire, and to keep the Japanese from slipping up on our positions. Before stopping, we had cleared a section along our front out to about seventy-five yards in front of our gun positions.

We had been firing artillery support for the Eighth Marines when just before midnight the cane fields came alive. No one was sneaking. You could hear the Japanese commander yelling to his troops, getting them lined up for an assault on our artillery gun positions. I was worried; it sounded like an army from all the noise and rustling being made in the dry sugar cane field. Everyone braced for the imminent banzai attack. We were like the Indians, waiting for the cavalry to come charging in. I was sent to my foxhole and prepared to defend our gun section. I still had two magazines with ten rounds in each one. I had a desolate feeling that there was an unequal balance of power, judging from all the preparatory noise going on in the cane field. I was hoping for the word to pull back, but looking around at the men who were all set for the onslaught, it was quite evident we would not give up our position. Not having much ammo, I got my carbine bayonet ready as added security. I was prepared for the worst.

We didn't have long to wait. The attack came fast with loud screaming, firing, and bodies crashing through the sugar cane. All guns commenced firing, accented by small arms and machine gun fire. Dirt and coral rocks landed on me. This was caused by all the bullets being fired, or the blast from the guns. I managed to fire all twenty rounds when the Japanese charged out of the cane fields into the area we had worked so hard to clear. I remained in my foxhole until I heard someone yell, "Cease fire." I got some more ammunition and remained on watch for the rest of the night. No one was certain this would be the end of the banzai attack. I expected another wave, or another attack from the survivors of the first attack. When daylight came there were approximately 150 dead Japs scattered along the front of our positions. Many had fallen on the edge of the cleared area of the sugar cane field. I checked the bodies in front of my position to see if any of the soldiers had been killed by a carbine. From the looks of the bodies, most of them appeared to have been killed by our two .50-caliber machine guns. These guns were located on the flanks of the battery, and had caught the charging Japanese in a devastating cross fire. We were very lucky not to have had any casualties in this encounter. This was one of those nights when I didn't think I'd be around when the sun came up.

A long trench was bulldozed out, and the machine pushed the bodies into the trench and covered them over.

The war went on. We moved to a new position and started firing our guns nonstop. I was put on a working party to keep our battery supplied with 75-mm ammo.

A cold rain started coming down in torrents, like a monsoon rain which soon saturated everything. I never though a rain could be this cold in the tropics. I shook with chills, and my teeth chattered. I was numb. I believe I came close to freezing to death. Being so cold, and with the lack of sleep, I became a walking zombie. I got to the point where I couldn't think for myself. I had to be told everything, and sometimes had to be given a shove to get me moving. I remember working in a daze. I would pick up a cloverleaf of 75-mm ammo from the pit and, slipping and sliding up from the pit, load it on the trucks, then slide back down into the pit for another cloverleaf. The sergeant in charge of the working party sent me back to the battery since I was of little use to him. When I got back to the gun section, I was immediately put on watch to spell one of the men so he could get some rest. I was relieved after being on watch for four hours. I went to my foxhole, which was half full of water. With my poncho wrapped around me I sat down in the water, leaned back, and went sound asleep. When I came out of my foxhole, my skin was wrinkled, and it was hard to walk with the deep wrinkles on my soaked feet. I felt better, and the rain had stopped. I had lived to see the sun come up one more time.

The guns continued firing support to the front lines. I was assigned to patrol the area in and around our gun positions. While on one of these patrols, we came to a small village of about five or six houses. The houses were still intact, but there were a few dead Japanese soldiers lying around the area. Some of the men chased down and captured a couple of chickens and took them back to our gun positions. They were put in a can and boiled. C-ration was added to make chicken soup. It smelled delicious, but as much as I hated C-ration I could not bring myself to taste the soup. I wasn't certain, but I was quite sure the chickens had been feeding on the dead bodies. Knowing this stifled my appetite.

I remained with the gun section until the island was declared secured. There was still resistance on parts of the island, and mop-up operations would continue. I can't remember if I volunteered, or if someone volunteered me for this mission. One gun section was sent to aid the Eights Marines in their mop-up operations, by delivering direct fire on the Japanese holdouts. On our way up to the front lines, we went up a road on the side of a mountain. Along the top of this hill were large guns overlooking Tinian town and Tinian Harbor. This town had the only good beach on the island, and the Japanese expected us to make our assault on this beach. The beach was heavily mined and barricaded, and with the large guns on the hill overlooking the harbor was like a big booby trap, waiting for the marines to land in the trap. This never happened. The Second Marine Division landed on the opposite side of the island on a beachhead about twelve feet wide. This unorthodox landing saved the marines from a costly battle if they had landed on the beach of Tinian town. Whoever made the brilliant decision to send a whole combat division of marines through such a small opening saved many lives.

After passing many of the big guns on top of the hill, we came to a wooded area. From that point on, we would have to disassemble the guns and carry the component parts in piece by piece. The howitzers were designed to be broken down and carried on the backs of pack animals. We didn't have any mules or jacks. We had to muscle them in by manpower. It was both awkward and clumsy getting the parts through a winding wooded trail. There was only one way in and out, and you had to have a guide show you the way. The whole area had been booby trapped with TNT and grenades. The Eighth Marines were set up on the edge of a 120-foot cliff about 100 yards from where we unloaded the guns. Once we got the guns, ammo, and all of our accessories toted in, we reassembled the gun on top of a large coral outcropping on the edge of the cliff. About 700 yards out from our cliff was another cliff that dropped down to the rocky shoreline. This cliff and rocky shoreline were controlled by about 200 Japanese soldiers. This enemy cliff made a wide arc, and where it curved you could observe the mouths of some of these caves. Our mission was to neutralize these caves.

Sgt. Schott was the section chief in charge, and he had us assemble the gun on this jagged coral rock with the muzzle of the gun pointing down to the caves. We could not use the rear trail of the gun to take up the recoil because the rear trail was pointing up in the air. We removed this trail and used ropes and sandbags to anchor the gun to the cliff. When we were ready to fire, I was to be the number one man, the man who pulled the lanyard, firing the gun. The first time we fired, the gun bucked like a mule and sandbags went flying over the cliff. I grabbed onto the front trail to keep from falling off the cliff. We retied the ropes and again anchored the gun. This time we used a lesser charge, which proved to be satisfactory to some extent. If we made much of a shift in deflection, we would have to untie the gun, make the shift, and again go through the routine of anchoring the howitzer. While we were firing the guns, we were exposed to the Japanese fire, but their small arms fire was at an extreme range. I ended up with a souvenir while on this cliff—a .31-caliber slug struck and lodged in my first aid packet. A U.S. Navy photographer took a picture of the gun in this position. The picture appeared in the book *Battle Stations*, published in 1946 by Wm. H. Wise and Co. of New York.

It was on that cliff that I ran into my boot camp buddy, Corporal Campo, from Detroit, Michigan. He was a squad leader in *I* Company and made combat patrols daily. When we weren't firing the gun, I started making patrols with Campo and his squad. On the first patrol we were fired on. I never saw a Jap, but the squad managed to kill two Jap soldiers. On one of these patrols we went down our cliff and patrolled out to the edge of the cliff controlled by the Japanese. We would lie down and fire at the Japanese over the edge of the cliff. The Japanese fired back at us from below. They would dodge in and around the huge boulders on the shoreline. You had to snapshoot at them as they tried to better their positions. While firing down at the Japs, one marine next to me fell backwards. I thought he was hit. I picked him up, and he had a gaping bullet hole in his helmet. He had a mosquito net waded up in the top of his helmet. The bullet penetrated the helmet and the net, missing his head by a fraction of an inch. Once he saw what had happened, he went into shock and turned white as a sheet. He was still shaking and very pale when we arrived back on the top of our cliff.

Any time you went on patrol you had to climb down a long rope to reach the bottom of the cliff. On another patrol, we went down the rope and about 800 yards from the base of our cliff to the enemy's cliff. As soon as we looked over the ledge, the Japs started firing at us. I snapshot any time I saw movement. At times they would dive or dodge to a new position, and I would get off a few good shots, but once they dove behind the big rocks and boulders, I was unable to tell if I had made a hit. We had quite a fire fight going. Some marines tossed grenades over the cliff. Cpl. Campo had given me two grenades, but I didn't want to waste them tossing them over the cliff. To my left was a marine with a Browning automatic rifle, or BAR, as we called it. This weapon was awkward, and the BAR man stood up to better fire his weapon. He had just started firing when he was hit and almost fell on top of me. At about the same time, two Eighth Marines guarding our rear opened fire. I thought they were firing at me. Dust, dirt, and rocks went flying all over me. I was still clinging to the ground when the two marines ran up to a clump of bushes that were located just to the left of the BAR man. They pulled out two badly shot up Japanese soldiers. They spotted the Japs when they fired point blank at the BAR man. The BAR man had been shot through the chest, and the bullet almost severed his arm on exiting his body. We got him on a stretcher, and I helped to carry him back. The bleeding was very bad, but he was still conscious and mumbling on the way back to our cliff. A cold rain started, and water and blood were pouring off the stretcher. I put my poncho over the man to protect him from the chilling rain. We had one devil of a time getting him up the cliff and trying to keep the stretcher level. After they had taken him back to the aid station, I started shaking and my teeth started chattering so bad that I thought they would break. I crawled under the old dirty and oily tarp that we had over the gun; this was the only protection I could find. After this operation, I had to pay about seven dollars for the poncho because I didn't have it at our first inspection. I didn't think the man would live. I heard later that he would live, but would lose his arm.

The Third Battalion, Eighth Marines had booby traps set in the bushes and shrubbery behind us to prevent the Japanese from charging in against us, trapping us against the steep cliff. One night when a booby trap exploded, the machine guns opened up, spraying the area with fire. The next morning we found three Japanese bodies in the brush; we never knew if there had been others who had chickened out and fled.

I thought we had a good defense set up to our rear, but one evening a lone goat wandered into our camp. Somehow the goat had meandered down the winding trail that wasn't booby trapped. Every marine was thinking barbecue, and started yelling and running to catch the goat. Once spooked, it took off in a straight line right through the bushes, tripping and exploding every booby trap in its path. I was sitting up high near the gun and caught a glimpse of the goat as he cleared the perimeter of our area. He was still running and didn't appear to be injured. The Eighth Marines had to remake and reset all the booby traps, placing some of them higher and at different angles. If the Japs had come charging in fast and low, they could have made it all the way to our gun position, trapping us against the cliff.

Early one morning, Maj. Chamberland, the battalion commander of the Eighth Marines, was scanning the front with his field glasses. He said he could see some soldiers just inside one of the caves. We already knew the range to the cave, so we loaded and fired. The round exploded just inside the mouth of the cave. The major said, "Well, they're not there now." He turned and went back to the CP.

A short time after this, one of our artillery officers came up to the front to investigate a report that artillery men were being used for combat infantry patrols. After the officer finished talking with the section chief, I was ordered to remain with the gun section. I was barred from joining any more patrols. Cpl. Campo continued making combat patrols. All I could to was have him a cup of coffee ready on his return.

Two Japanese men surrendered to the Eighth Marines. They came up under a white flag. They represented a bunch of civilians being held as hostages by the Japanese soldiers. We gave them candy and cigarettes to take back to the civilians, as a token, to let them know we wouldn't harm them.

At about this time a destroyer arrived in the bay. This destroyer had been requested to fire into the caves where our gun was unable to reach. The ship was in a hurry to leave, but agreed to wait one hour for the two civilians to return. The destroyer waited an additional fifteen minutes before it started firing the five-inch anti-aircraft guns. Shortly after it started firing, one of the civilians came running back toward our cliff. Once we got him up on the rope, we learned the Japs had caught them and had taken away the candy and cigarettes. They shot one man and threw him into the water. They were going to shoot this man, but when the destroyer started firing the Japanese dispersed, giving this man the opportunity to escape. The ship continued to fire into the cave until it received a hit on the main deck from a Jap knee mortar. This assistance was over. The destroyer left the bay post haste, never to return.

Before this part of the island was secured, the Eighth Marines had rescued over 200 men, women, and children from the Japanese soldiers. After this mop-up operation was over, we disassembled the gun for our return to the battery. It was on our about this time we were informed that the battle for the island of Guam had been secured, and Guam was now in the hands of Americans.

When we returned to the battery, it was set up around an old frame house. The CO was using this house for the CP office and living quarters. The rest of the battery had pitched pup tents, or shelter halves. Everyone was waiting for the rear echelon to catch up with our six-man tents.

We had received a bunch of new replacements, and as usual they complained about everything. They hated the food, the guard duty, the insects, and the toilet facilities.

We spent much of our time digging the guns in and getting them set up in a defensive position, just in case the Japanese tried to retake the island.

While we were on the front, the battery had been sending out patrols every day around the area. I heard complaints that there was no reason to send out patrols because the island had already been secured. A week after returning to the battery, I was assigned to one of these patrols. We were moving along the ridge line of a tree-covered ridge when the corporal in charge told me to go down and patrol along the base of the hill. It was very hard to get through the thick underbrush. I had to duck and weave to try and plow my way through the underbrush. At times I had to back up and try a different route. In one place I ducked a low limb and caught a faint glimpse of a movement to my left. I froze and waited, but I couldn't see or hear anything.

All I could see to my left was a wall of shrubbery. Still standing in the same tracks I again detected the movement. I began to move my body forward and back while at the same time looking out of the corner of my eye rather than straight on. I finally spotted a dugout of some kind about twenty yards to my left. I worked my way around until I could look into the dugout. I could see three rifles, grenades, a green marine expeditionary water can, some explosives, and a roll of primer cord. There was a hole going down into the base of the dugout. The movement I had seen was a head being jerked back into this hole. I called to them in Japanese not to be afraid and to come out.

I waited and only one small soldier came crawling out of the hole. I pointed and motioned for the others to come out. The soldier indicated to me that he was the only one, and he seemed very anxious to leave. I still had the two grenades given to me by Campo. I took one of these and was getting ready to pull the pin. The soldier, seeing what I was going to do, motioned for me to stop. He started talking to someone in the hole. Soon three more soldiers emerged from the hole. One was a giant of a man. My carbine started feeling like a toothpick. I yelled for the patrol leader to send down some help, and fast. A number of men responded to my call. Two of the new recruits and one older man were detailed to take them back to camp.

I remember one of the new men was named Duncan. He came up to me when I got back to camp and told me that a news photographer had taken their picture with the prisoners, and it would be published in his hometown paper. He said, "Since I got all the glory, I wanted to give you this as a souvenir." He handed me an Omega watch that he had taken off one of the prisoners. I never searched the prisoners for weapons; all I did was watch their hands. If they had moved toward the weapons or grenades I would have fired. I kept this watch for years, until it finally got away from me. It was two days before I got a chance to get away from camp, go back to the dugout, and retrieve the Jap rifles. I went to the same area and thought I could go directly to the dugout. This became a big puzzle to me. I spent the entire day going up and down the hill, but I could not find the dugout. I was sweaty, tired, and worn out when I returned to camp. I had also been stung by yellow jackets. I never went back to this area. Something happened; maybe the Japanese took everything and covered the place up. I'll never know.

From the top of this ridge line, you could look off in the distance and see a couple of old houses. I thought about going over to the houses to look around, but we had been warned more than once to stay away from the houses, as most of them contained booby traps. I never went to check them out, but three men in our battery didn't heed this warning. A few days later they went out to look for souvenirs. While they were looking around the houses, one man set off a booby trap that killed him instantly. The CO wrote to his wife. I have no idea what he said, but I'm sure he spared her the truth—he was killed due to his own misconduct.

The rear echelon finally caught up with us, and we were able to pitch our six-man tents. This was a lot better than living out of pup tents. While we were in this position, most of the old men who had contacted malaria on Guadalcanal began coming down with the malady again. The men would be sent to the field hospital, and then to the States.

Bill Moss was a good marine. He would come down with chills and high fever, go to the hospital for a week or so, and for some reason be sent back to the battery. One evening he returned from the hospital and was told he had guard duty that night. A cocky new lieutenant to the battery had decided we would have inspection of the guard. All the guard would be expected to be clean-shaven and have a clean rifle and dungarees. Moss didn't have time to prepare for this inspection. He was still weak from being in the hospital. He fell out for inspection with a stubble of beard, a rusty rifle, and dirty dungarees. He looked like a character out of Bill Maudlin's cartoons of Willie and Joe. Moss was standing next to me in the second rank. The cocky lieutenant started the inspection of the first rank. While inspecting the rifles, he asked each man a question on sports, general orders, or some other topic. He inspected my rifle and asked me my eighth general order. He stepped in front of Moss, and you could see the surprised expression on the lieutenant's face as he took the rusty rifle. He asked Moss his name, and then asked him who was the Secretary of the Navy. Like a Bill Maudlin character he gave the surprise answer, "I don't know sir, I'm not interested." This brought a laugh from some of the men, including me. The lieutenant almost dropped the rifle, turned, and handed the rifle to the sergeant with orders to dismiss the guard.

Later, Moss was summoned to the CO's quarters. He asked Moss what the problem was, and Moss told the CO, "All the men who served with me on Guadalcanal have been sent home. They sent me to the hospital for a while, and back to the battery. On my first day back, I'm put on guard duty. I don't think it's right." The CO gave Moss a case of beer and told him he would not have to stand guard duty that night. After the CO talked to battalion headquarters, Moss was ordered to pack up. He was leaving the following morning. We had a farewell party that night for Moss. Taps sounded, and the party continued. The executive officer came in and said we were making too much noise. He told us to keep it down, and we could go on with the party. He turned before going out of the tent and said, "I'm not easy; I think you men deserve a break." Moss boarded a jeep the following morning on the first leg of his journey home. As far as I know Moss was the last man who had contracted malaria on Guadalcanal to leave Fox Battery.

I had stopped writing to the girl in Burbank as I thought it would be a waste of time. I didn't think I would live to see the States again. Anyway the censors blocked out any good or bad news you wanted to share with someone. About all you could say was that you were well and still alive.

Seeing all the burr-headed recruits running around reminded me that I was no longer a recruit replacement. I was one of the old men.

There was no liberty, and no place to go if you were granted liberty. We stayed busy by using scrap lumber and some bags of Japanese cement to build a mess hall. We had it screened in so the flies wouldn't compete for the food. We were still getting C rations, but they came in one-gallon cans. This was supplemented by the long cucumbers growing in a field next to the mess hall. Once in a while we would receive a different ration, but eventually always reverted back to the standard C ration.

It seemed like we had been there for ages, but at last we received orders to pack up; we were leaving Tinian.

We arrived back on the island of Saipan, and debarked near Tanapag Harbor. We moved to the north end of the island to set up our camp a short distance from an air strip. This whole area would be used for the Second Marine Division camp. It covered quite a large area.

We got our tents set up and spent many hours hauling and pouring crushed coral rock on the ground inside the tents. This was to be our floors. We dug trenches around the outside of the tent to prevent a muddy floor. We had cots to sleep on with no mattress. We had one blanket and a mosquito net to go over the cot. I still had to wear my socks at night, or the mosquitoes would drive me crazy, feeding on the bottom of my feet where they touched the netting.

A division bakery was set up, and bread was distributed to all the units. This bread was always full of weevils. If you picked them out of your bread, there was nothing left. I couldn't bring myself to eat it, and I couldn't stand to watch the other men eat it.

An amphitheater was set up in a valley, with sand bags to sit on stacked along the side of the valley. You had to walk two miles from our battery to get to the theater. Much of the time it rained, or the film broke. If neither of these things happened, the generator ran out of gas, or stopped working altogether.

Once we got our camp in order, we tried various things to keep in shape. We had a lot of gun drill, and took the guns out on maneuvers to fire them. Sometimes we were interrupted by someone spotting Japanese soldiers, and then we spent most of the day patrolling the terrain. We took a large number of hikes and organized a track team and a football team. I participated in anything that came along.

Six months on this island, and we were still getting C ration. The army and the Seabees were getting vegetables and fresh fruit. It appeared the marines were on the bottom end of the totem pole when it came to material and food distribution.

While in this camp we had about twenty men crack up or suffer from battle fatigue. Depression and worry about the next campaign were the biggest culprits. I, too, was becoming bored and depressed. Early one morning a Jap zero livened things up by strafing our camp. Our tent flaps were rolled up, and I was looking right at the nose of the zero and could see the trail of the machine gun tracers coming straight at us as they came in at treetop level. Several other men and I ended up in the muddy drainage ditches that we had dug around the tents. Luckily no one else was injured in this surprise visit.

I didn't want to crack up, so I had to entertain myself. I left one Saturday to visit the B-29 air base. I wanted to see if they would let me buy something from their PX. I had heard that they had a well-stocked PX. All our PX ever had was a good supply of sardines, crackers, Baby Ruth candy bars, and cigarettes. I found the air base PX easily enough, but they wouldn't sell me anything. I was told if they started this that they would be swamped with marines. As I came out of the PX I met and talked to one of the airman of a B-29 bombing crew. He invited me back to his tent to meet the rest of the crew. I also met the colonel commander of the B-29 and was very surprised to find him quartered with the enlisted men. I spent the whole evening there; they treated me like royalty. I made a number of weekend visits to see and talk to these men. I brought them Japanese rifles, bayonets, and anything else I could lay my hands on. They bought me things from the PX which I would take back and sell, share, or trade to the other marines. We always had a nice visit every Sunday evening; they treated me like one of the crew. One Sunday evening I went for the usual visit and found two lieutenants in the tent taking inventory of the contents. The officers told me that the crew had been in a bombing raid over Japan. The plane had been hit and crashed at sea; there were no survivors. This was devastating news to me. I never thought anything would ever happen to these men. I never went back to the air base, and I never got acquainted with any other men from the air force.

One day, after walking over a mile to the beach for a salt water bath, I decided to go south and look around. I was on the west side of the island and went south until the beach played out, and a rocky shoreline became too rough to travel. I was just looking around and enjoying the scenery. In one place I came across a dead marine. There wasn't much left of his body. Everything was rotted. His rifle, bayonet, and parts of his cartridge belt were still with him. His helmet was a hard piece of rust. If there had been a dog tag, it was long gone. The bayonet was too rusty to put on the rifle, so I stuck the muzzle of the rifle in the ground and put the rusty helmet on top of the rifle stock. The wood from the stock had survived the elements much better than the metals had.

After this I continued following the terrain of the coast south as the land continued to rise. I found myself on a very high cliff, so steep I was afraid to get real close to the edge and look down. I was somewhere in the vicinity of Marpi Point, where the Japanese had thrown themselves over the cliff rather than surrender to the marines. I went inland about 100 yards and continued on south for some distance when I came to a huge crevasse. It was very wide where it dropped down the steep cliff to the rocky shoreline, but it gradually tapered off as it went back inland. It became nothing but a wide crack in the earth small enough to jump over. I walked up and looked down. It appeared to be a bottomless pit. I was still too scared to get close to the rim. I laid down and looked over the edge. After my eyes became accustomed to the darkness, I could make out a well-beaten path leading back from the water to where I was looking down. The path disappeared back under the ground; how far it went, I'll never know. I'll bet it was a good hiding place for a lot of Jap soldiers. this was quite a spectacle, but I didn't dare try exploring this by myself. In leaving the area I went east and ended up on the road that went south to Tanapag Harbor. It was getting late, so I hiked north and back to camp. There had been some fatalities along this road, as nurses and servicemen were still being sniped at as they traveled after dark.

I had the midnight shift on guard duty one night, and instead of trying to get some sleep, I played cards until time to go on watch. I was marched up to the motor pool on the side of a hill and put on watch. The guard I relieved passed his orders on to me and then joined the formation. The corporal of the guard marched them back down the hill. It was a very cold night. I was wearing extra clothing and my field jacket. Right after I had

47

been posted I heard a noise, and was surprised to see the guard being marched back up the hill. I halted them and asked the corporal why he had returned to this post. He told me it was time for me to be relieved. I told him that I had just been posted. He asked me what time I came on. I told him twelve o'clock. He said it was after four, and to pass the orders to the new sentry. This was the shortest guard duty I had ever stood in the Marine Corps. I had spent four hours standing on my feet in the same place, sound asleep. I usually was very dutiful when it came to guard duty. I was determined not to let this happen again.

Without the bombing crew to visit on weekends, I started being lonely, especially on weekends. I was beginning to worry about myself, and decided the best way to contain worry was to keep moving, so I decided to try exploring the island some more.

I started out one Sunday going east from our camp, past the motor pool on my right and the gun park on my left. I continued down this small road with no particular goal in mind. I had explored all I wanted to west and south of our camp. I always carried my carbine and my Marine Corps K-bar, the marine excuse for a trench knife, on these excursions. I never wore a helmet, and I never took any grenades with me. I continued on this road for over a mile until I came to an intersection of another small road going north and south. This was a tee intersection, with my road dead-ending. Looking east I could see a large hill. I decided to climb this hill and look around. It took almost two hours to reach the top. Once on top I could see the air base to the north, and looking west I saw the marine camp that extended almost to the ocean. Looking south I could see nothing but hills, with the mountain in the background. Looking east I could see that I was overlooking a dead-end valley. The hill I was on extended around this valley in the shape of a large horseshoe. I was in the middle of the bend. I could see smoke rising as I looked down into the valley.

I went down the hill into the valley to investigate the smoke. I took advantage of what cover I could find as I neared the camp. There were caves and dugouts on both sides of the valley, and I was afraid a guard might be posted in one of these caves. As I approached the smoke, I could smell the odor of rotting rice. I got close enough to see the camp, which had only one fire with something cooking over it. There were marine water cans and other stuff scattered around. Leaning against a log were four .31-caliber rifles. I picked up two of these rifles, made haste back down the valley to the dead end, and then climbed the hill. It seemed like it took me forever, but I finally reached the top. I climbed in a zigzag manner which made it easier, but I was still soaking wet when I reached the top. I sat and scanned the valley for quite some time, but I never detected any movement in the valley. These rifles were loaded. I was so anxious to get away from the Jap camp that it never occurred to me to unload the other two rifles. It was just as well, as any delay could have been deadly.

My market for souvenirs had faded, but I found a new market in the recruits who took the rifles off my hands. They insisted I give them the ammo. I thought nothing of this, so I gave them the ammo to go with the rifles. A couple of days later, an Eighth Marine patrol escorted the two recruits into our camp and turned them over to the first sergeant. These two men had been in the vicinity of the Eighth Marine area and started test firing the rifles against a hill. The Eighth Marines were very sensitive to the sound of Japanese weapons and dispatched a patrol hoping to catch some Japs. The two recruits were very lucky the Eighth Marines were not trigger-happy. The two men received a scolding and had the rifles confiscated. A massive shakedown of the battery to eliminate any other dangerous weapons ensued. They took four rolls of film that I had taken of the bombardment of Tarawa, but they didn't find my grenades.

Our drinking water was brought in and poured into lister bags. These were round canvas bags with spigots around the bottom. The bag was suspended by three poles set up like a tripod. You were not allowed to get water out of these bags in anything but your canteens. I was only shaving once a week, and it seemed silly for me to go out and get water in my canteen, then come back and pour it into the steel part of my helmet, which I used as a pan. Early one morning I ran out to the lister bag and poured a little water into my helmet. I turned around to leave and almost knocked the gunnery sergeant over. As punishment, I spent a number of evenings digging a hole ten feet deep to be used for an outhouse. Every evening the men would wave to me on their way to the movies. I generally dug until taps sounded. Someone scrounged up some lumber, and I helped to build a good four-hole outhouse over the pit. I did such a splendid job on this project that the officers had me dig a hole in their area and construct them a four-hole outhouse. My popularity was growing.

We had been on this island for almost a year, and the food we ate was coming from the same old C rations as the ones we ate in combat. We had a Quonset hut set up for a mess hall. About all the cooks did was heat C rations. Our officers were served the same food.

One of the men in the Sixth Marines was so distraught that he wrote all of his complaints in a letter and sent it to his mother. The censor, instead of censoring the letter, added a note to the bottom saying, "What this man says is true; we officers are getting the same thing." Somehow this letter ended up in the office of the Commandant of the Marine Corps. One evening the Sixth Marines had a surprise visitor, the commandant himself. He borrowed some mess gear and went through the Sixth Marine chow line, took it over, and dumped it into the garbage can. After inspecting the camp, he departed. We found out later the lumber that was supposed to be used to build our tent decks had been diverted for use by the commanding general to build a nice house on the side of a hill overlooking the ocean. The general was given the opportunity of buying the house or facing disciplinary action. He agreed to buy the house. I thought he was getting off easy, but a short time after the commandant's return to Washington, the general received his transfer orders to Guam. I don't know whatever happened to the house. Some officers of the Seventh Field Depot were relieved of their duties because of this incident. After the inspection by the commandant we received lumber for our tent decks, and we started receiving fruits and vegetables. This was quite a boost to the batteries' morale.

The time many dreaded had arrived. We had orders to prepare the guns and all the accessories for combat readiness. Once we were ready to move out, we had orders to tie the tent flaps down, as we would be returning to this camp after the next campaign. It occurred to me at the time that the area would have to have constant guard around the area to keep it secured from the Japs who were still holding out in the hills around us.

I couldn't believe my eyes when I saw the changes that had taken place on this island as we traveled south to embark on a ship docked near Tanapag Harbor. Buildings had been constructed all over the island, even a big hospital. It was hard to visualize any place or area as it had been when we first landed.

Everyone worried about the battle to come. Most of the men were feeling as I, that after so many encounters and battles, you feel you have used up all of your odds and the next fight will claim you for good.

I hated being aboard ship, so with nothing to read I started sketching marines on deck and in the mess hall. It was a good way to pass time, and the men got a kick out of it. I gave most of the sketches away.

A few days out we began our briefing on the battle plans for the island of Okinawa. This was a big island, and wouldn't be easy to take. We were told that we would not land, but rendezvous off the southern end of Okinawa and then start to shore. This feint at a landing was to draw as many enemy troops as possible to the southern end, while the landing of the main assault forces would take place in the center of the island. We were up for our usual two o'clock breakfast. This was standard for all D-day landings. Boats were circling all over the surface of the water, getting organized to start for shore. It was a nice morning when we got underway for the beach, which was still being bombarded by air and naval gunfire. We were in line, ready to make a run for the beach. The beach itself looked safe enough, and I thought at the moment we were making a big mistake by not going ahead and making the landing. As we neared the beach, we started drawing fire. Artillery rounds were landing in the water as the wave of boats made a big U-turn and headed for open water.

All of the other boats were going through this same maneuver before returning to the ships. The fake landing had been completed.

We boarded ship and learned that the assault forces had made a successful landing in the center of the island, and the battle for the island was now in progress. We raised anchor, departed the waters of Okinawa, and headed back to the island of Saipan. I was disappointed not to have landed. It was like sitting on the sideline while a football game is in progress. You want to get in on the action. We returned to our old camp on Saipan. Our mission had been completed.

On our return, our camp on Saipan looked as if it had been hit by a tornado. It was in shambles. The Japanese had come out of the hills for a visit, and probably used a sword or machete to cut all the ropes supporting out tents. It looked as if a high wind had blown the tents all over the area. Our mess hall and reefer had been vandalized, causing extensive damage. We had to start rebuilding the camp all over again. While we were rebuilding the camp, the Second Marine Division made a sweep of the whole island, and killed or captured over 200 Japanese still holding out. This was a fast sweep, so I know they didn't get them all.

We had just about completed getting our camp back to normal when we received orders to prepare to move out. This time we were a part of the Eighth Marine combat team. Our mission was to deliver 75-mm artillery support for the Third Battalion, Eighth Marines. A few days at sea, and we began our briefing. We were to land on the small island of Iheya-Shima, located just to the north of Okinawa, and secure this island so it could be used as a radar warning station for the forces in and around Okinawa.

After the usual two o'clock breakfast, we boarded the boats while the island received air and naval bombardment. This was a small operation, so the bombardment wasn't as intense as it generally was in similar operations. There was no opposition on landing, but a couple of marines were killed or wounded from misdirected friendly fire. We had the island secured in three days. Even without opposition, we still had to check every part of the island for possible enemy troops. We spent a miserable twenty-four hours out of these three days in a cold, drenching rain. My teeth chattered, and I had chills from the icy rain. This time I was glad to be aboard ship; I was tired, dirty, and completely worn out.

Again we could not drink the island water because it contained parasites. If you smashed one on your skin while washing, it caused small blisters to form on the skin. I'm not sure what would happen if you accidentally drank the water. I wasn't about to be the one to find out.

I thought we were returning to Saipan, but once aboard ship, we received a change in orders. We were to relieve the Seventh Marines on the southern part of Okinawa. The Seventh Marines had lost most of their men and had pulled back, dug in, and were trying to hold their position. The Eighth Marines under Col. Wallace were to relieve what was left of the Seventh.

When we first boarded this ship, I ran into Cpl. Campo, my boot buddy from Detroit. He was extremely depressed, and he said he had a feeling that he would be killed in the next action. I did my best to cheer him up, but he said he just knew he would be killed. I couldn't get his mind off that thought. He asked me to talk to the battery commander and see if he could be able to transfer from I-3-8 to F-2-10. He was so despondent that I agreed to talk to the battery commander. I explained to the captain exactly why Campo wanted to transfer out of his company. The captain told me that the records were back on Saipan, and he could not effect a transfer without them. He did ask me to remind him when we returned to Saipan, and he would do what he could to get Campo transferred into Fox Battery. I relayed this information to Campo, but he was still insistent that it would be too late; he would already be dead. After our skirmish on the island of Iheya Shima, I thought Campo would be all right, that his fears were imaginary, but after we received the word that we were going to land on Okinawa, he was more distraught than ever.

We landed just north of the town of Naha. Everything was blacked out as we got our trucks and guns ashore. We had moved a short distance from the beach to a staging area before moving out in a convoy. While waiting, a Japanese Betty bomber came over and dropped a load of bombs in and around the dock area. I dove into the nearest depression, which happened to be a deep muddy rut left by one of the trucks. It was quite a scare, and I was a muddy mess with no way to clean up. We continued on south through the town of Naha. We finally got to a pre-determined area and got the guns set up and ready to give artillery support to the Eighth Marines.

The Seventh Marines had lost most of their men going up a valley. The valley was the only way through a heavy ridge line, lined with steep cliffs. The Seventh Marines only had a few men left and were unable to rescue their killed and wounded.

Col. Wallace, our commanding officer, decided to have his Eighth Marines scale the cliffs and pull the machine guns up to support the other troops scaling the face of the cliffs. The first day the Eighth Marines had advanced 300 yards beyond the ridge line. Once the Eighth Marines were on the move, we kept busy firing the guns and delivering artillery support for the infantry.

While we were firing our guns in this position, we were aggravated by a lone Japanese soldier with a Namboo light machine gun. Every evening just about sundown, he would lie in the mouth of a huge cave about 500 yards in front of our guns and spray his fire all over our positions, making everyone dive for cover. This was quite a nuisance, and continued for about five days before a squad of marines were sent out to dispatch who we had been referring to as machine gun Charlie. From our position we could observe the squad of marines closing in on the mouth of the cave. They were coming in from both directions, and were going to toss a satchel charge inside to neutralize the cave. A satchel charge is about ten or twelve blocks of TNT tied up in a burlap bag. Once the fuse inside is triggered, there is no way to disarm it before it detonates. The marine patrol came up cautiously to the mouth of the cave and tossed the satchel charge inside. Wow! The whole face of the hill exploded. Rocks rained down all around us. Some as big as cars landed just in front of our positions. This patrol eliminated the machine gunner, but they had sacrificed themselves to complete the mission. The Japanese generally kept the floor of the cave covered with blocks of picric acid, an explosive similar to our TNT. When this ignited, it detonated the ammunition stored in the cave.

One morning we received word that all of the enemy's heavy artillery had been knocked out, that it would be strictly an infantry operation from that point on. We got busy, and cleaned and put the covers on the guns, but we still were to remain in our present position until the island had been secured. We could still fire the guns on a moment's notice, if needed.

Since there were no duties to be performed in the gun section, I decided to go up to the front lines and try to find a sniper's rifle for a souvenir. This would be my fourth campaign, and I was sure I would go home after this action. I caught a ride and thought I was going to the front lines, but the truck was going to some other unit behind the lines. I got off the truck and started walking down a side road. I came to a small village and started looking around. It was evident that the Eighth Marines had already been through this area because of all the sandbags and foxholes. I came to a small town bank that had the side of the building blown out. There was money scattered all over the floor. I took an empty sandbag and filled it with various denominations of Japanese bills. During this trip I didn't see any civilians or Japs. Once back at F-2-10, I took one bill of each, and the other men did the same. There were still a lot of them left over, and the next morning we used some of the bills to start a fire; other bills blew out across the ground.

The next night someone heard a noise in front of our position. Everyone was alerted, and the machine guns on our flanks started firing in a criss-cross pattern. The next morning three dead Japanese soldiers and one Japanese woman were found. They had tried to slip through our position to get to the rear.

Right behind our position was a natural spring. I washed and shaved and put on some wrinkled khaki that I had been carrying in my pack. It felt good to be out of my dirty dungarees. Another man in the section did the same. His name—I'm not sure of the spelling—was Terrashaveitz; we called him Ski. He wanted to go to the front lines with me. We both borrowed a .45-caliber pistol for protection and set out for the front lines. We caught a ride on a truck and told the driver we were going to the front and to let us off at the nearest place to the front lines. He stopped on a main road that curved off to the right and told us the front lines were over that way, pointing up a small road leading south.

Ski and I started off along this road. To our right was a sandstone cliff that started about knee-high and continued to rise up as we went along the road to a height of about 100 feet. There was a valley or ravine that dropped down to our left. This road wound around a hill in a wide arc. We came to an army jeep sitting on the road while two army soldiers huddled in a small cutout in the sandstone cliff. We asked them for a ride, and they gave us a funny look as they said, "Jump in the back," and both came running out and jumped in the jeep. Just as the driver was getting ready to start the engine, Japanese rifle fire started kicking up the dust all over the road and cliff. We jumped out of the jeep and hit the deck, fast; dust from the bullets was so thick you could hardly see. We had no sooner hit the deck than we were up and running for the protection of the small sandstone dugout. The soldiers said they had been pinned down for over an hour.

Once the Japs ceased firing, the soldiers told us that before only a couple of riflemen had been firing at them; now it seemed like a platoon of riflemen. After waiting for some time, and having had time to catch my breath, I told the two soldiers that since we were wearing khaki and pistols, that the Japs probably thought we were officers. I made the remark that if they got in the jeep without us, they could probably drive off without any shots being fired. I believed the Japs would save their ammunition for us. The two soldiers gave this some thought, then got on their mark like two Olympic sprinters, ran the ten yards to the jeep, and sped off in a cloud of dust. If any shots were fired, I didn't hear them.

Ski and I were now stranded. He made some remark about what a dumb thing to do. We only had our two pistols, but we checked and got them ready just in case the Japs sent out a patrol to kill or capture us. I don't believe they were interested in taking prisoners this late in the battle for the island. We waited for some time, but no patrols approached the dugout. We could peek out from this cutout and see across a wide ravine or gully to a mesa with small trees and scrub brush. This mesa was on about the same level as our road and extended about 700 yards away to a sloping hill. We believed the rifle fire was coming from this hill, but we couldn't be sure.

It was getting late when we finally made the decision to run back down the road for about 100 yards to where the sandstone cliff became small enough for us to jump over for protection. We got on the mark, just like the two soldiers, and took off back down the road running like two scared jackrabbits. Ski was in front of me as we made tracks. Bullets started kicking up dust and making loud thuds and splat noises as they slapped against the sandstone cliff. Gravel, rocks, and dust were scattered all over me as I continued to run. It seemed the faster I ran the heavier the fire. The intensity became so great that I turned and made a mad dash back to the protection of the dugout. As I was doing this high-stepping boondocker ballet, I thought Ski had probably reached safety by now, and if I had continued to run forward I would have reached safety too. I went through a thick cloud of dust and made a headlong dive into the small cutout of the cliff, only to have Ski land right behind me. He must have turned back a fraction of a second after I did. We were both out of breath and drenching wet. It sounded like an army firing at us. I'm sure the Japanese thought we were officers, and consequently had brought more firepower to bear on us. Being wet from the exertion of the run, we decided to wait until it got dark before attempting another run.

It was almost dark when we heard a roar from a tank coming around the road. It was a marine tank going back down the road. I waved to the marine who had his head sticking up in the gun turret, and the tank stopped. I asked the marine for a ride. He said, "What the hell are you guys doing up here in khaki?" I told him that we were looking for the front lines. He pointed across the ravine and said, "They're over there." I told him we had been pinned down and couldn't get out. He had us get on the side of the tank and gave us a ride back down to the main road. The marine in the tank said that this road was a spur that circled out and around the hill. All movement on this road was exposed to the front lines. We started back north and finally got a ride back to camp. It had been one tiring and exciting day for both of us.

When we returned to F-2-10, the battery was still sitting in the same position. We were told the island should be secured at any time.

I was put on guard duty that night and had plenty of time to think about our excursion to the front. It was very dangerous, but at the same time, you had a sense of adventure and excitement that is impossible to describe. The Eighth Marines were moving fast so I decided to go back in the morning to the winding spur of the road and explore the area that had been held by the Japanese. I was quite certain the Japanese soldiers would be gone from this area, as the Eighth Marines were continuing their push southward.

This would be my last chance to try and get a sniper's rifle for a souvenir, as I expected this to be my last campaign. I asked Ski if he wanted to go. I don't remember what he said; either he had guard duty, or he didn't care to go.

I set out early the next morning all by myself. This time I was dressed in dungarees with my carbine, K-bar, and my two hand grenades. Getting a ride toward the south end of the island was easy, as there seemed to be a lot of traffic going back and forth. No vehicles were turning off on the winding road along the sandstone cliff. I left the main road on foot and was again walking along this winding road. About halfway between the cutout in the sandstone cliff and the main road, I began to get a queasy feeling in the pit of my stomach. I was exposed to anyone on the hill and mesa off to the left, and across the ravine.

Every time I took a step I expected the crackling of rifle fire. I had only one consolation; a lone marine in dungarees was no big prize. I was still gun-shy remembering the day before when we had been strafed by small arms fire. I paused in the small dugout of the sandstone cliff and scanned the area from which the ground fire had originated. I left the road, went down into the ravine or dry wash, and up the other side. Once I was up on the flat plateau or mesa, I looked back and saw that I was level with the road. This plateau gradually rose to a hill or knoll in the distance. This was probably where the Japanese had been located when they started firing on us the day before. I started for this knoll, going through bushes, tall grass, and small scrub trees, some in clumps so thick you had to walk around them. The tallest trees were on the knoll about 600 or 700 yards from the road, which had been my estimate. About halfway to the hill, I came across a bloated body. It was swollen to twice its normal size. I started to pass on, thinking it to be a Japanese soldier, when I noticed a dark blot on his breast pocket. This is the Marine Corps emblem. On a closer examination I discovered his dog tags. They were messy, but I got them off and put them in my pocket and continued toward the knoll in the distance. I had estimated the distance from the road to the hill as about 600 to 700 yards. I believe it was closer to one-half mile. This is probably why the Japs sprayed the road with fire rather than taking aim. The road would have been at maximum range for small arms, even their long .31-caliber rifles.

I was almost to the hill when a loud shot rang out; it was very close. I hit the deck and rolled for cover, crawled, and dodged until I was hidden from the hill. Utilizing what cover I could find, I worked my way up closer to the knoll. The sniper who fired was using a .25-caliber rifle. It had about the range of my carbine. I decided to get this rifle for a souvenir. I worried my way closer to the hill. Most of the hill was barren of trees. There were only a couple of trees on the top that you couldn't see through. They had enough foliage on them to conceal a sniper. The hill was like a tripod, composed of three small ridge lines leading up and converging at the peak. I was on one of these small ridge lines; another one led down from the hill and off to my right. The third ridge line was off to my left but mostly to the rear of the hill. I couldn't observe it completely. I got into position, took careful aim, and fired three shots into the trees at the top. The sniper almost fell out of the tree and started running fast down the ridge line to my right. I was trying to get a good shot at him when he dove into a bunker on the face of the ridge. I had him.

I could watch the entrance to the small bunker as I crossed over the depression between the two ridges without going to the top of the hill. I went above the bunker and approached it very slowly. I decided to throw a hand grenade inside, then go in, retrieve the rifle and leave. The bunker resembled a midwestern storm cellar. I pulled the pin on my grenade, and was ready to toss it into the dugout, when I remembered the squad of marines being blown up a few days before. I got up close to the bunker, placed my left hand above the entrance, and looked down into the interior of the bunker. I could only see the floor of the bunker, but this was enough. Scattered all over the dirt floor of the bunker were loose yellow bricks of picric acid. I backed off as fast as I could. My hands were shaking so bad. I had a hard time trying to put the pin back into the grenade. All the while I thought the Jap would come out shooting. I was still above, and still watching the mouth of the bunker, when I finally got the pin back into the grenade and reattached it to my belt.

I decided to go up on the hill, find a place where I could sit, and wait for the Jap to come out. I was almost to the top of the hill when three four, maybe more, shots rang out. I bolted just as the sniper had done. I ran off the ridge line and across the depression to the other side of the ridge line in order to go back the way I had come. I heard more shots as I got to the bottom. I ran like a deer, dodging behind shrubs and bushes, keeping some cover to my back. I passed somewhere near the dead marine, but didn't see him. I was doing the boondocker ballet, making tracks as fast as my feet would move. I was almost, or a little over halfway back to the road when I stopped.

I was wringing wet and out of breath. I took a sip out of my canteen, and thought no one would pursue a lone marine. I no sooner had the thought when more shots rang out. I didn't know if they were coming from behind me, or firing from the hill. I took off again, and didn't stop until I came to the rim of the ravine or dry gully. I went down into the gully, but I was afraid to start the climb up the other side to the road. If I did this, I would be a sitting target, that is, if the Japs were in close pursuit. Instead of climbing up the side of the hill, I turned and ran up this dry gully back toward the main road. Staying in the ravine offered some protection. I got to the end of the gully and made my climb up to the main road. I looked back across the plateau, trying to detect any movement of the Japs, but I couldn't see any sign of them. I had lost one of my hand grenades along the way while doing the boondocker ballet. I caught a ride back to F-2-10. It was getting late and I was chilly, still being wet with sweat. Once I got some dry clothes on I was back to normal. In spite of a good scare, it had been an exciting day. While I was on my skirmish to the front, F-2-10 had had our battery picture taken. I saw the picture; it appears everyone was in it except me.

I turned the dog tags I had found on the dead marine over to the CO. I never did get a look at the name. The tags were filthy and would have to he cleaned before you would be able to read the lettering. I told the CO I had found the tags on a dead marine. I didn't tell him where. This ended my souvenir hunting on Okinawa.

The following day after my trip to the front, General Buckner was killed on a tour to the front lines. He wanted to see the Eighth Marines in action. He was killed while visiting the Third Battalion, Eighth Marines command post.

A few days after General Buckner was killed, the Third Battalion, Eighth pulled back and bivouacked about a mile from our position. Mop-up operations continued, but the island had been declared secured. I decided to hike over to I-3-8 and visit my buddy, Campo. Once I got to the Eighth Marine area, I was directed to the battalion command post. On entering the CP tent I was greeted by the battalion sergeant major. I asked for directions to I Company. He wanted to know who I was looking for. I told him, "Cpl. Campo." He asked if I was a relative. I said, "No, we are boot camp buddies." The sergeant major said, "Cpl. Campo was killed yesterday evening on hill 310. He was shot through the head by a sniper." I asked for his address so I could write to his folks, but the sergeant major told me not to worry, that the CO would write to his family. I will never forget Cpl. Campo for having such a strong premonition that he would be killed in this action. I just wished I had been able to get him transferred into our battery. I had lost a good friend and a good marine.

A few days after my visit to the Eighth Marines, we loaded up and again embarked for our camp on the island of Saipan. The battle for Okinawa was now history. It had been a bloody battle, but at last the island was in the hands of the United States.

When we arrived back at our camp on Saipan, I was very surprised to see all of our tents still standing. Either the holdout Japs were fading out, or our forces had maintained good security in our absence. Nothing in camp had been damaged. We were now back at our old training routine in order to ready ourselves for the next campaign. I heard nothing about anyone being returned to the States. We still had a war to fight.

I had received a letter from home giving me the name of a marine unit that a cousin of mine was assigned to. The unit, a 90-mm anti-aircraft artillery battery, was located on the island of Tinian. I received permission for a couple of days' leave to visit my cousin, George N. Skelton. I caught a boat over to the island and had a devil of a time trying to find his unit. The island was nothing but B-29 air strips covering most of the island. You had to travel miles to get on the other side of an air strip. It was late in the evening when I found the unit. They had their guns set up very close to where I had captured the four Japanese soldiers. It was unbelievable to see the change in terrain. The hills were gone; there was nothing but flat B-29 concrete runways extending as far as the eye could see.

I had a nice two-day visit and was ready to get back to my unit. This in itself took a lot of doing, as none of the roads had signs. It seemed that everyone who gave me a ride was going the wrong way. It took almost a whole day to get back to the docks to board the boat for Saipan. I vowed this day would be my final visit to this island.

While the battle for Iwo Jima was raging, we received word to board ship. We embarked for Iwo Jima. The floating reserved had been committed to the battle. We were now the floating reserve. Before we reached Iwo Jima, we received word that the island had been secured. We returned, debarked, and returned to our camp.

Almost any time we took the guns out to fire them, we would come across signs of Japanese holdouts, and we would spend the rest of the day hunting Japs. The holdout Japs didn't seem to be causing any trouble, but they still posed a threat. Everyone carried his weapon and ammo every time he left the camp area. The mosquitoes had diminished and didn't cause us as much discomfort, but I still had to sleep with my socks on. Every evening a C-47 flew low over the island spraying DDT. This had been going on for just about a year. The numerous big flies had been reduced from massive swarms to almost nothing.

I finally got a letter from my brother. After an extensive stay in the hospital, he was assigned to limited duty in Kingsville, Texas. The many operations had restored his face, but he had lost the sight of his left eye. I also received news from home. I had learned that many of my neighbors and school buddies who had gone into the service had been killed or wounded. Some of the others who didn't enter the service had either died in a car accident, from drug overdosing, or from committing suicide. A few were serving time in prison for various crimes.

The 75-mm pack howitzer was declared obsolete. It was phased out of the Marine Corps. In place of the 75s, we received 105-mm howitzers, which were larger guns. We stayed very busy training and learning these guns.

The island of Iwo Jima was being used as a fighter support base. the B-29 bombers would leave the Marianas, and the fighters with smaller fuel tanks would join them off Iwo Jima and escort them to and from Japan. This eliminated many losses of the B-29 bombers and their crews. It came too late for my friends in the B-29 that had crashed at sea.

Sgt. Schumate was made our gunnery sergeant, and we were still training on the 105-mm howitzers when we received news that Japan had surrendered. The company street exploded; at first I thought it was mail call because everyone ran excitedly out of their tents. I was very happy. It had been a long time. I had survived. Why? I'll never know. The war was over after two and a half years, which seemed like a lifetime to me. I should now be returning to the States. Instead of preparing our guns for a stateside ride, we started preparing them for combat. We had the battery ready to move on a moment's notice. We didn't have long to wait; we were going to Japan to be a part of the occupying forces. We struck camp, and this time we took our tents. It seemed like I had spent most of my life on this island. I had had enough. I hoped this time that we were leaving for good.

The Second Marine Division cemetery looked real nice, with all the crosses in neat rows. I always wondered how they ever got the crosses over the right bodies, when most of the dead had been given a mass burial.

I watched Saipan fade in the distance over the phosphorescent wake of the ship, and breathed a silent prayer for all the hundreds of marines who remained behind and who would be a part of the island forever. May they rest in peace.

Japan

Once aboard ship, it was hopeless to even think about going home. I had to busy myself with something to keep my mind off of depressing thoughts. I had lost everyone that I had ever been close to. Even the older men had given me a feeling of loss and isolation when they had returned to the States. About all I had to do was read and sketch the marines and sailors. Even this was not easy. The sailors were always washing down the decks, chipping paint, or repainting. I will always hear the echoes, "Hey marine, you can't sit there; hey marine, you'll have to move forward, or go below deck." This was an endless routine, but still it was the only way the sailors could perform their duties. It was so crowded, and the marines were always in their way.

We finally received a briefing on our proposed landing in Japan. We would follow the scheduled invasion plans that had been formulated by the high command. We would make our landings on the beach or shore of the town of Nagasaki. No one knew what kind of reception awaited. We would go in loaded with live ammo, just like with any other invasion, and be prepared to fight if we happened to meet any opposition on the beach.

Our ship entered the mouth of the bay leading into the docks of the city of Nagasaki. The mouth of the bay narrowed to a channel as we neared the town. I was very tense and braced myself. Large guns from the hills on both sides of the channel looked down on us. Thank goodness they did not fire. If Japan had not surrendered, and this was the initial invasion plan for the Second Marine Division, we would never have made it to the beach. The Japanese had enough big guns to blast us out of the bay.

We dropped anchor and went down the cargo nets into the waiting boats. Everyone had his weapon loaded and expected some kind of opposition. We landed without any problems. A large number of Japanese had gathered on the beach to watch us, mostly out of curiosity. I didn't see any of them in uniform.

I was given a squad of men and ordered to remain on the beach until all the equipment and supplies had been unloaded from our two ships. This was the most responsible assignment that had ever been entrusted to me since I had been in the Marine Corps. I was determined to do a good job. I didn't have enough men to keep up with the amount of supplies coming in. Boats were lined up waiting for us even though we were working our butts off. Some of these boats were loaded with 50-gallon drums of fuel. There were only three pieces of machinery on the beach called cherry pickers, capable of lifting these drums out of the boats and onto the beach. These small cranes were being used by officers. Every time I tried to get one, I was told I would have to wait. The officers had rank over me, so all I could do was let the boats wait. An officer next to me wanted me to supervise his working party while he was gone.

I was wearing a field jacket with no stripes, and I guess this officer thought I was an officer. I took over more docks and starting moving more supplies and equipment using the two working parties. I had to keep moving and yelling louder. It was getting very late when the beach master came by and said he had to leave, and I was to take over the entire beach; he had already notified the other working parties. A short time after the beach master was gone, I pulled all three cherry pickers off the other docks, and put them busy unloading the fuel drums from our boats. The lieutenants along the beach began bitching, but by then my boats had been cleared. I told my men we would be there for two days moving our supplies, one piece at a time from the boats fifty yards upon the beach. I told my men to give the Japanese men watching us some cigarettes, and see if they would help us unload the boats. We soon had a long line of Japanese men passing supplies and stacking them high upon the beach. By two o'clock all of our boats had been unloaded. I told one of the officers I had to leave, and he was to take over until the beach master returned. As I was departing with my working party, I heard the officer yelling for the cherry picker to get down to a certain dock.

Most of the battery had departed. I was taken out and put on watch at the edge of the atomic bomb area, to keep all United States vehicles from passing through this area. I was called many names that night by all ranks of officers and gentlemen who wanted to use this road for a short-cut. I refused passage, so they had to take the long way around. I was relieved from this post at eight in the morning. When it got light enough, I could see the devastated area of the atomic bomb site. There was nothing but rubble as far as you could see.

We boarded a truck in Nagasaki and went about sixty miles north to the small town of Isahaya. On the way we passed through miles of tunnels equipped with machine shops and other types of machinery that were used to turn out war materials. None of these facilities had received any damage from the bombings of Nagasaki or the surrounding areas. The capability of the Japanese to manufacture arms and munitions was still intact. I was worn out when we arrived in this camp. I laid down and went to sleep. It was almost dark when I woke up. There were only a few men in the barracks, when one of the men returning from chow came in and said a Japanese man was out in back of the barrack and was asking to see someone. I went outside. A Japanese man with a pretty young girl was waiting. He wanted to sell her for 600 yen. I checked my wallet to see what kind of money I had saved from the damaged bank on Okinawa. I had one 1,000 yen note, plenty of money to buy the girl. The girl was the man's daughter. Since we had orders not to fraternize with the Japanese, we had to decline his offer. I thought this was an awful thing to do, but later I learned this was a normal transaction in Japan.

After we got our barracks cleaned up, we started making trips to the surrounding towns and disarming them. One building was designated as an armory, and all the arms were turned into this building.

In my spare time I would sit out on the taxiway of the airstrip and sketch Japanese zero fighter planes and gliders that were parked along-side the runway. Most of these planes had been cannibalized for spare parts or were in a state of total disrepair. It was doubtful any of them could fly. I was often interrupted by marines, coming out to fill their Zippo lighters with the high-octane aircraft fuel. One evening I noticed five marines working on the zero fighter planes. I asked them what they were doing. They said they had pumped fuel into one of the planes and were going to try and get it started. They added, "If we get it started, we are going to fly it to the States." They were planning to stop on the islands along the way to refuel. These men were so homesick you could not reason with them. I was hoping they wouldn't get the plane started, especially since none of them knew how to fly. Some of these men were from the motor pool, and did know something about mechanics. They worked tirelessly every evening for about five days, and cranked on the engine before giving it up. I've seen men this homesick in book camp who would try and go to Houston and back on a three-day pass from San Diego.

Another quiet evening while sketching, I was interrupted by three men from the motor pool. They were in a jeep, and told me they were going to launch one of the gliders. One of the men said that with a good wind blowing across the water, he believed he could take it all the way to the States. They checked out one glider and found it in good condition, with the exception of the release mechanism. They solved this problem; they would use the release mechanism of the trailer hitch on the jeep to release the cable by splicing two lengths of cable together, and then towing it behind the jeep to get it airborne, and then release the cable from the trailer hitch on the jeep. This hitch was two-half-moon sections that locked together. It was generally used to pull guns or T-4 carts behind the jeep. These marines were determined to the point of being crazy. I hadn't seen anything like this since I was seven years old.

When I was seven years old, I used to ride with my dad, once a month, twenty-four miles into the town of Terrel, Texas to buy groceries. This was a long, slow, and tiresome ride in an old wagon behind two mules. It took a good twenty-four hours to complete the trip. As we neared the town of Terrel, we would pass by the insane asylum. This institution had long, tall fences around the grounds. There were men along the fence digging with shovels to get out. Once they got too deep, the guard would move them to a new location, and the digging would start all over again. This kept the inmates busy and out of trouble. They were always working when we passed—digging and dreaming of freedom.

After a lot of work the three marines had the glider ready for towing. I set aside my sketch book to watch them take off. I was still doubtful that they would ever get it off the ground. They got everything lined up on the runway. The cable looked to be much too long as the jeep started moving down the runway at a fast clip. About halfway down the runway the glider lifted off and went up like a kite. For some reason, probably from disuse and salt water, the jeep's towing hitch would not release. One man was bent down over the back of the jeep, frantically striking the hitch with a tire tool or pry bar trying to free the cable. The jeep was out of runway, and went all the way to the fence before coming to a sudden stop. The glider continued in the air and over the fence. When it reached the length of the cable, it jerked the jeep up and into the fence, spilling the men out. The glider made a wide arc before it came crashing down, landing on one wing and the nose. At the same time the jeep from the slack in the cable fell off the fence. The pilot of the glider was uninjured, but the driver of the jeep had a broken arm. The jeep didn't seem to be damaged as they drove the man with the broken arm to sick bay in it.

The following evening I went out to sketch as usual but was stopped at the gate by a sentry on duty. He told me the airfield was restricted, and no one was allowed on the airfield without special orders from the battalion commander. This put an end to my sketching sessions.

I made liberty every weekend by hiking into the little town of Isahaya. The first time I went into town I had my photograph taken in wrinkled khaki; this was the best uniform I could find. It had been in my backpack for quite some time. After I had my picture taken, I visited the world bank and found out that the Japanese currency I had found on Okinawa was good as gold. It could be changed into currencies of other countries, including the United States. All the money I had given away and discarded on Okinawa would have made me a rich man. I did enjoy a number of liberties utilizing what Japanese money I had in my wallet.

It wasn't hard to spot the geisha houses in town. On every street that had one, a big sign had been erected that read, "Out of Bounds to All Military Personnel." Anyone entering these houses had to remove his shoes and leave them by the front door or on the porch. When the officer of the day made the rounds, he would pick up the shoes, and any marine returning to the base barefooted would generally end up with some kind of punishment. This kind of information spreads fast through a marine camp. When you visit one of these houses, take your shoes with you. So far I was staying clear of these houses because I was due to go to the States at any time, and I didn't want to jeopardize my leaving by getting into any kind of trouble.

One evening while on my way back to the base from a liberty in town, I passed an out-of-bounds sign; looking past this sign was a big geisha house about a block away. I wondered what the women in this house looked like. I passed a big church, or similar building, and noticed the back was partly fenced and enclosed with hedges. It reminded me of a church back in Hawaii. I went to the rear of the building and found a narrow alley that led down to the narrow street. The sidewalk leading to the geisha house was just across the street. I was contemplating visiting this house, because the only time I would show myself in an out-of-bounds area was to cross the small street.

While I was pondering the chances of being caught, two Japanese women appeared on the porch and waved me over. I was still uncertain, and while I was waiting the women sent a beautiful Japanese girl over in a nice clean kimono. She took my hand and walked me to the house. Once inside, I was given a choice of several women. They all looked a little old to me. I indicated to them that I wanted the pretty girl who had been sent outside to get me. I was told I couldn't have her. I started to leave, and suddenly they stopped me and had a long talk with the pretty girl. Soon it was agreed that I could go to the room with her, but one of the other women would be standing outside the door. I agreed, and a nice romance started to bloom. Every weekend I would visit her and take her candy, cigarettes, and other things. She was extremely pretty and looked like most of the pictures you see of pretty geisha girls on posters and calendars. After about four visits, she would laugh and smile at me, which was unusual for the girls who worked in these houses. We had fun and enjoyed each other's company. I always stayed until one of the older women told me I had to leave. I always looked forward to the weekends. About the sixth or seventh visit, I was told that she was gone, and i would have to select another girl. I refused and departed. About two weeks later

I went back to the house. Again I was told she wasn't there, and wouldn't be coming back. I had no way of knowing what happened. I didn't know if she was gone, or if they just didn't want me to see her. We both knew our romance wouldn't last, but it ended much too soon for me. I really missed seeing her. She was a doll. This ended my romantic adventures in Japan. I never visited another geisha house.

While I was visiting the Japanese girl in the geisha house, I learned that the health department inspected the girls once a month. I thought this was a good idea, until I found out that the inspection was from the waist up.

There was a point system set up to determine who would be returned to the States first. I had the third highest points of anyone in our unit, so I was sure I would be going home soon. Instead of being sent home I was sent down to a camp on the outskirts of Nagasaki and began training in the gun section all over again. This seemed like such a waste to me because I was sure I would be leaving at any time. However I heard nothing of going home. Instead I started standing guard duty on merchant marine ships and the Chinese embassy. This was rough guard duty, with six hours on and six hours off. The most demanding post physically was the entrance to the Chinese embassy. You spent most of your duty at attention, or present arms, because of all the officers going in and out of the embassy at all hours of the day and night.

I had been in the camp in Nagasaki for over a month when I was ordered to pack my gear. I was to leave the following morning. A few other marines and I were taken down to the train station early the following morning and put on board a train bound for Sasebo, Japan. This was a busy station, and the train was packed to capacity as we pulled out of the station. What a big disappointment as I thought we would be boarding a ship bound for the States. We were now on board a primitive Toonerville trolley. The windows were open, and the sulphur fumes were stifling. This little train would start up a mountain and continue slowing down to a crawl. You could get out and walk as fast as the cars were moving. Once on top of a hill or mountain it would gain momentum and race down at full speed, only to be slowed on the next ascent. This was a very slow and tiring process. In the meantime the train stopped, loaded, and unloaded passengers at every little station along the way. It seemed like it took ages, but at last we arrived in Sasebo, Japan. The marine camp was just outside of the town. We were informed that no ship was available, and no one knew how long we would be there. There would be no liberty, but we would stand guard duty, pick out a bunk, and be prepared for an indeterminate stay.

On my first guard duty I was driven way up onto a big mountain and placed on watch at the entrance of a large manufacturing complex. This was also left untouched by the bombing raids over Japan. I had a small one-room guard house, with an old coal stove for heat, at the entrance. Since this post was located so far from the camp, I was assigned twenty-four hours on duty and twenty-four off. The small coal stove kept the place warm enough, but the fumes from the stove were awful. I had many visitors while on this post; they were all Japanese who had never seen a real white Caucasian up close. They were well-meaning, and they brought me all kinds of fruits and vegetables. I was afraid to eat anything they gave me. The Japanese used human excrement as fertilizer on everything. Anyone not accustomed to this kind of food came down with acute dysentery. After twenty-four hours on this post, I was always happy to get off the mountain and back to camp. I continued guard on this mountain for over three weeks.

I requested liberty more than once, but it was always denied. A couple of marines sneaked out of camp and went into the small town of Sasebo. They went into a geisha house and were having a good time until the officer of the day entered the front. The two men ran out of the back door. One made it back to camp, but they caught the other one. It was dark behind the house, and he had run off a ledge into a gravel pit, breaking his arm. After that I gave up trying to get out; the camp was sealed up tighter than ever. I could never understand why we were restricted to this camp. It would have been nice to know, but no one ever told us.

I had just about given up hope of ever returning to the States, when one of the sergeants came into the barracks and began calling out names, including mine. We were instructed to pack up and be ready to board ship the following morning. I was already packed and had been for a long time. We embarked the next morning in a cold heavy fog. At long last we were finally bound for the States. This would be a fast trip, and we didn't have to worry about blackouts, kamikaze planes, or submarines blowing the ship out of the water. It was almost unbelievable; at last we were under-way. Many times I had thought I would never live to see that day. It gave me a good feeling that I was going home. It seemed I had spent most of my life in the Pacific.

The phosphorescent wake of the ship was dissipating at about the same time as the fog and the odor of the bay. I was leaving Japan, and would never come this way again. The Japanese people were nice, and I would have loved duty in this country under different circumstances, but for now it was better to leave. The Japanese had been my enemy for a long time. I could never trust them, awake or asleep. They had left an indelible stamp on me. Maybe in time I'll forgive, but I'll never forget.

Farewell to the sea gulls, the raunchy bays, and farewell to the memory of a beautiful Japanese girl, as beautiful as a princess, who gave my life and hers a little meaning and happiness for a very short period of time.

A part of my life always seems to be ending, but at the same time another path emerges, and you following this, not knowing where it will lead you. Like a bad dream, it's beyond your power to alter the course.

On our way back across the Pacific, I occasionally caught a glimpse of an island and wondered what island I was seeing. Just about every piece of land in the Pacific has marines buried on it. It made me wonder again how I had survived and others did not. I'll never know why. I always went where I was sent—to the front lines, or rear—it never really mattered to me. I never complained, and I was never a sick bay soldier. This is what we called the men who continually lined up at sick bay every morning to try and get out of work or combat duty.

I was going home with no close buddies to talk to. I didn't have a close buddy, so I spent most of my time reading or on the bow of the ship, watching the flying fish and riding the bow up and down. The men with seasickness continued to think I was crazy for doing this.

The ship made good time. The Pacific fell behind, and I was glad. I didn't like it, and I hoped I would never see it again. I was going back into a void. I had no plans ahead once I was discharged.

Stateside and Home 1945

We docked in San Diego after what I thought was a very fast trip across the wide Pacific Ocean.

We were billeted in Camp Pendleton after getting our sea bags which had been stored away when we left the States so many months ago. Everything in the sea bag was musty-smelling.

It was Christmas Eve, 1945, and everyone was restricted to the base except the men who had relatives living in the general area. I still had the address of Gladys Ponessa, so I turned in this name as being that of my sister. I wasn't intending to see her, but I was using her as an excuse to get off the base. We hadn't received any pay, and I only had three dollars. On Christmas Eve I got dressed to leave. I had no emblems or ribbons, and my wool marine greens smelled like mildew. I wore my old dirty boondockers. I had tried to polish them, but they were almost worn out and wouldn't take a shine. A recruit fresh out of boot camp presented a better image of a marine than I did, but I was determined to get off the base, regardless of how I looked.

The same old road was still lined with marines trying to hitch a ride. I walked past them, as I had done so many times before, and started hiking the eighteen miles to the main gate. Cars loaded with marines passed me regularly; none of them stopped. I didn't care; I was in no hurry. It was Christmas, and I was happy to have made it back to the States in one piece. The morning was cool and brisk, and I enjoyed the stateside smells along the road. I was relaxed and didn't have to worry about anyone sniping at me. The outdoors smelled of home, and the evergreens added a special smell to emphasize this day as Christmas Eve. I counted my blessings; the Lord had been kind to me, and I was extremely grateful.

I was almost to the main gate when a car stopped and offered me a ride to Los Angeles. With no destination in mind, I accepted the ride. The driver dropped me off at the Greyhound bus station. I drank a cup of coffee while I tried to decide what to do. I started walking around, looking at the sights, when I decided to walk to Burbank and visit Gladys. It had been so long since I had written to her that I didn't know if she was still single, or if she still lived at the same address. At least it gave me an objective. I walked most of the night, still happy to be back in civilization. It was early morning when I had arrived in Hollywood, and most of the businesses were closed. I was surprised to find a bowling alley open, and it was even stranger to find people bowling on some of the lanes. I had a cup of coffee, sat in the spectators' seats behind the players, and catnapped to daylight.

It was a chilly morning as I resumed my hike toward Burbank. I arrived in Burbank about eleven o'clock. After asking directions, I arrived at the correct address at about noon. After hiking all this distance, I was still hesitant about knocking on the door to the house. I had only a few dates with Gladys before going overseas, but I had never met her parents, and didn't know what kind of reception I would receive from them. A grey-haired, portly man answered the door. I asked for Gladys; he invited me in and asked me to sit down. He told me Gladys would be out soon. I had never met this man before, and I kept thinking he had me mixed up with someone else. While I was waiting, he brought me a big mug of beer. This really tasted good after the long hike. Mrs. Ponessa and Gladys's twin sisters came in and introduced themselves. I was surprised that both of them had black hair. Gladys was a blonde. When she entered the room, she looked beautiful and was just as friendly as I had remembered her. The family invited me into a spacious dining room where a big turkey dinner was waiting. I believe it was the best turkey dinner I've ever eaten. It had been over twenty-four hours since I had eaten, so naturally it was delicious.

After dinner, Mr. Ponessa gave Gladys the car keys and some money. We took in a movie and had a nice visit. She was still single and still look-ing. Although she was sweet and very pretty, I wasn't in love with her. Anyway in my present position, it would have been impossible to make any kind of commitment. After the movie she drove me to the bus station in Los Angeles and, over my protest, bought me a ticket back to Ocean-side. I begged her not to, but she said her father had given her the money and had instructed her to see me on the bus. She sat and waited with me until it was time to board. We gave each other a big hug and kiss. We both knew this was for the last time. This was a Jewish family who had treated me like a member of the family. To this day, a Christmas never goes by without me remembering them with warm feelings and hoping all of them still enjoy a nice Christmas. These people have always, and will continue to, hold a special place in my heart.

I got off the bus north of Oceanside and hiked the eighteen miles back to the barracks, arriving early in the morning, just in time for roll call. Christmas day of 1945 had been a happy day for me. I was hoping the next Christmas would be spent at home.

After everything had settled down after the holidays, we started our processing for discharge.

We were marched to one barrack and were asked if we had been injured or had suffered any disabilities while in the service. I told them I had hurt my back on Saipan, and had ringing in my ears from all the artillery rounds we had fired. I was told if I registered a claim I would be sent to the naval hospital in San Diego for thirty days to have my claim evaluated. It had been a long time, I was determined to go home, and I didn't intend to put it off for thirty days in a hospital.

We were marched into another hall for a reenlistment lecture. The sergeant invited everyone in and said the smoking lamp was lit. All the smokers lit up and continued smoking during the reenlistment speech. While the sergeant was giving the gung-ho talk, the smokers were looking for a place to douse their cigarettes. The big sergeant noticed this and told everyone it was all right to stomp the butts out on the floor. This brought a laugh from the crowd as most marines don't throw cigarette butts on the floor. Without any ash trays the marines had no choice but to mash their cigarettes out on the floor. After the completion of the lecture, the sergeant said he had reenlistment papers for anyone interested in signing up. Everyone roared with laughter, but no one went forward to reenlist. After a short wait the sergeant said, "Very well, everyone pick up your cigarette butts and move out." This completed the reenlistment part of the processing.

Most of the men made reservations for a flight home. I decided I had had about as many close calls as the law of averages allows. I didn't dare tempt fate by flying home. Four of us marines paid a driver fifty dollars each to drive us home. He was taking anyone going to cities along the way from Camp Pendleton to Memphis, Tennessee. By switching drivers along the way we made very good time.

The driver let me off in front of the Jefferson Hotel in downtown Dallas about midnight. I caught a streetcar going to the Oakcliff section of Dallas. I was going to transfer to the Trinity Heights car once we crossed over the Trinity River into Oakcliff. When we got to the transfer point the last car to Trinity Heights had just departed. I rode on into Oakcliff and got off at Jefferson and Ewing Avenues. There was no traffic at this time of night. I was over eight miles from home. All of the businesses were closed, and I had no way of making a phone call. My folks didn't have a phone, so the only alternative was to walk home in the middle of the night carrying my sea bag.

I had only been back to Dallas for a couple of short visits since my departure in 1941. I had been in trouble a number of time with the juvenile authorities in 1940 and 1941. My juvenile probation officer was Mr. Barrow. He was Clyde Barrow's father. I thought it very strange to find a father of a desperado to be such a nice man and a probation officer. He told me if the authorities caught me one more time they would keep me. Rather than go to the Gatesville reform school, I decided to leave Dallas. Here I was in Dallas again, but I was way past being a juvenile.

I walked south on Ewing, and when I neared the Marseilles Park bridge, I opened my sea bag to eliminate some weight. The sea bag was full and very awkward to carry. To use Marine Corps terminology, I had to s__t can a number of heavy items. One was my overcoat. With my load a lot lighter, I continued on past the railroad tracks. The Black Street gang used to control this area. I had been away for five years, and if there was still a gang, they would be new, and I doubted if they would recognize me. However I didn't intend to take any unnecessary chances. Just before I came to Black Street I turned and went east up through a wooded area. I had lived in this neighborhood at one time and was very familiar with all the trails. I came out at the end of Claude Street. I was relieved not to have met anyone in the wooded section. This was one neighborhood you didn't want to be caught in, in the daytime. At this time of night it could prove disastrous. I tried to tiptoe through the middle of this street without disturbing anyone. I had made about three blocks when the dogs started barking. My sea bag was still too heavy, but I wasn't about to stop. I kept moving as fast as I could go. This was mostly a black area, and whites were not welcome.

After making it through this neighborhood, I was now in the Grant Street gang domain. It had been a long time, but I still didn't know how they would take my presence on the street. These were not organized gangs, but were older kids in certain neighborhoods who banded together and who would not hesitate to jump any stranger who didn't belong or live in the immediate area. I was just lucky this was a week night, and in the early morning hours. It had taken me almost four hours from the time I got off at Ewing and Jefferson Avenues in Oakcliff to make it to Corinth Street in Trinity Heights. I was worn out and soaking wet with sweat, but at last I had made it home. It was nice to be home. Everything was about the same as I remembered it. The house was a little crowded with two sisters and three brothers still living at home. My youngest brother was born right after the battle of Tarawa which is in the Gilbert Islands. They had named him Tony Gilbert after this battle. This was the first time I had laid eyes on him. Dad was still working at the Ford Motor Company. I had enjoyed the visit home, but somewhere along the way I had ceased being a kid. I was now a grown man, and I had to find a job and move out. I didn't want to be another burden to the family.

On one of our discharge briefings we were told that we could go to the unemployment office in our city and apply for unemployment benefits until we could find a job. The unemployment office in Dallas told me that as long as there was a job for which I qualified, I would be unable to collect benefits. They would give me the address of a possible employer and tell me they had an opening. I was sent to Oakcliff, then to South Dallas, and then to another address in North Dallas. Each time I filled out an application and was told they would notify me if I was selected for the job. My folks didn't have a telephone, so the only way I could be notified was by mail. The unemployment office never gave me more than one address at a time. I didn't own a car and I didn't know the bus schedule, so I spent many hours walking to find a particular address. After walking miles, and reporting back to the unemployment office, I would then be given another address at almost the same location from which I had just returned. I played their game for over a week before I finally gave up and left Dallas.

I went to Houston, thinking I would stand a better chance of landing a job in this city. Three days in Houston proved me wrong. This town had a surplus of men returning from the service. I spent three days in Houston before giving up the search for a job. I hated to go back to Dallas and live at home, but without a job I had no choice. I was on my way out of Houston when I passed a marine recruiting poster with a sign in front that read, "Reenlist, Pick Your Post of Duty." After inquiring, I found that since I had not been out of the service for more than thirty days, I would be able to reenlist and keep my rank. This would be the best thing to do until the job market settled down. I went back to Dallas, got my things in order, and reenlisted, picking Corpus Christi as my post of duty.

On arrival, and after checking into the naval air station in Corpus Christi, I was assigned to the marine guard detachment. The marine guard had open gate liberty. Anytime you were not on watch you could go on liberty.

My stay in Corpus was very short, but I'm going to relate this story. It's very odd and funny in some ways how I set out on a simple liberty in a town that I had never been in before, not knowing anyone, and became involved in the lives of other people in such a short time.

I went on liberty the second night I was in Corpus intending to go to a movie. To pass some time, I stopped in at the Roosevelt bar and ordered a beer. As I sat drinking my beer, an elderly woman sitting on a stool next to me kept staring at me. Finally she spoke, "Are you new in town?" I answered this in the affirmative, and she began telling me all about the town. While she was talking she ordered me another beer. I asked her what she was drinking. She said, "Warm milk, that's all I ever drink." She wanted to show me the town. Our first stop was a nice restaurant that featured spaghetti. We had an enjoyable dinner. The lady told me her name. It was a long French name, but she told me everyone called her Roxie. Every place we went, people called her by name. She paid for the meal and left a good tip for the waitress. Later we went to a movie, and she introduced me to everyone who worked in the place.

The following night I met her at the Driscal Hotel at her request. She wanted to take in a piano recital. While the recital was in progress, Roxie started criticizing the player in a voice loud enough to be heard by others. This, to me, was rather embarrassing. Two women sitting close to us invited her to go up and play. Roxie declined, but said that she only lived a couple of blocks away, and would gladly play for them if they wanted to come to the house. Her house was a huge two-story structure located about two blocks away on the same hill as the hotel. The inside of the house was expensively furnished with big oil paintings on the walls and pictures of Roxie seated in front of a piano when she was much younger. She said, "At one time I was the greatest pianist in Europe, but an accident damaged two of my fingers, and I stopped making public appearances years ago." She played for us, and the ladies agreed that Roxie was a better pianist than the lady who had given the recital. Her piano was magnificent. It was a very large and expensive instrument. I was not a musician but I, too, thought she played rather well.

Roxie invited me to supper the following night. She had a pretty girl she wanted me to meet. We were going to have supper and then go to an expensive night club. The following evening it came a downpour; streets were beginning to flood as I made it to Roxie's house. We were just sitting and talking, waiting for the girl, when she called and told us she would be unable to make it because of the weather. She had a pleasant voice on the phone, and we agreed to meet some other day. Roxie and I enjoyed a nice supper and had a long talk. She was engaged to be married to a sergeant who worked as a provost sergeant at the naval station brig. Before I departed I promised Roxie I would go over to the brig and meet her future husband. I could picture the sergeant as an older marine ready to retire from the service. This was the picture I had in mind when I entered the brig the next morning. I asked the sergeant at the desk, "Where can I find Sgt. Todd?" He said he was Sgt. Todd. Roxie had already told him I would be over to meet him. I was very surprised to see that he was about my age. He was very nice and showed me the new car and diamond ring that Roxie had given him. He said they were going to be married soon, and I was invited to the wedding. He told me she had already taken him down and introduced him to her bank president and hotel manager. I went back to the barracks with mixed feelings. I had no idea how long this romance had been going on. I didn't like it, but decided to keep quiet because it was none of my business.

I remained on base for the next few days, and spent a lot of time playing cards. One morning the clerk posted a list of names of the men who were being transferred. I was really surprised to see my name on the list. I immediately went to see the first sergeant and asked why I was being transferred. When I had reenlisted I had picked this station as my post of duty. The sergeant told me he had over two hundred men who had reenlisted and had picked this station as their post of duty. He was only authorized sixty men; he had orders to transfer all surplus men. A lot of men complained, but there was nothing we could do. The orders were final.

The night before leaving, I went into town to have supper with Roxie. She had cooked a nice farewell supper for me. While we were having supper, she asked me what I thought about her future husband. I told her the truth. I thought he was after her money and anything else he could get. I guess she really loved the sergeant; she said he wasn't like that at all. I thanked Roxie for showing me the town and making my liberties very special. As I left for the base, I gave her a big hug and wished her the best. Another curtain had descended on a short span of my life. I will never know how this story ended, but I hope it had a good one.

I had hoped for a weekend at home before leaving, but once back at the base all liberty was canceled because we would be leaving the following morning.

We boarded a troop train in Corpus Christi. Our destination was Port Royal, the town of Beaufort, South Carolina, and Parris Island, the Marine Corps boot camp for the eastern seaboard. This was an old train and had straight-backed wooden seats. They were packed so close together that you had to sit up straight with your knees crammed against the back of the seat in front of you. This was very awkward and tiresome. We were only given a sample of food, and everyone was starved when we pulled into the station in Montgomery, Alabama. The conductor told us we would be there for about thirty minutes. I got permission from the gunnery sergeant to run into the station to get some sandwiches. I took one man with me and ran back to the terminal into the restaurant. I bought what sandwiches were already made up, so there would be no delay. Coming out of the terminal with a large sack of sandwiches, we ran into another man who was sent to get us because the train was leaving. We ran as fast as we could, but we reached the end of the platform just in time to see the train disappearing around a distant curve.

We were stranded. I was stuck with thirty sandwiches that I had paid for. I also had neglected to bring my cap and jacket. The three of us went back to the terminal, and I called the Montgomery Air Force Base and explained our plight. They dispatched the MP jeep to pick us up. They took us to the base, assigned us to a bunk, issued us a chow pass, and gave us a free ticket to the movie. The next morning we were driven to the train depot and put on a civilian train. According to our tickets we would get off in Yemasee, South Carolina, and then catch a Greyhound bus to Beaufort and Parris Island. The civilian train was much nicer than the troop train, and we enjoyed the trip.

We debarked at two o'clock in the morning in the town of Yemasee, South Carolina, with below-freezing temperatures. The two marines with me had their uniforms on, but all I was wearing was my marine green wool pants and a light khaki shirt. We walked all over the town, but we couldn't find anything open. I was about to freeze. We finally came across a national guard armory. We pried open the back door to get inside. There were bunks inside, but no mattresses or blankets. I would sit on a bunk until I was almost numb, and then I would run up and down the aisle to keep my circulation going. The other two men were doing the same thing, as their jackets weren't enough protection from the cold. It was one miserable night. I was glad to see the sun come up the next morning. We found the bus stop and learned the first bus would come by at eight-thirty in the morning. This was another cold wait. When the bus finally arrived, we looked like three zombies boarding. It felt like heaven to be on a warm bus, but as soon as the feeling came back to my feet they hurt so bad I felt like screaming. The two toes that had been frozen in New Zealand gave me more pain than the rest of my feet. Beaufort was only fifty miles from Yemasee, but we had enough time to thaw out before leaving the bus. Getting to Beaufort was no problem, but trying to find our unit on Parris Island was something else. No one had reported us missing and, to my knowledge, the unit didn't have a name. We spent most of the day in the provost marshal's office while inquiries were made. Our unit was finally located. It was billeted in the boot camp barracks next to the rifle range. Gunnery Sgt. Taylor hoped we could find our way back and had not reported us missing. Parris Island is a big base, and the last place I would have looked was the rifle range. The gunny got chewed out severely for this, but that was the end of our ordeal. Parris Island was a miserable place. The sand fleas were worse than mosquitoes. They would eat you up. I felt so sorry for the boots in training. After watching them during some of their hectic inspections, I considered myself very lucky to have gone through my boot training in San Diego.

After one disappointing liberty in Beaufort I decided to remain on the base. The base had a good PX and a large bowling alley. The rest of my spare time I spent fishing. I would make some good catches, but threw them back because I didn't have any way to cook them.

Rifle training in the marines is hard work. No one liked the many hours on the snapping in range, but this practice was the backbone for becoming a good marksman and was an important part of the course. It was a puzzle to me why we were undergoing such extensive rifle training.

One day while on the firing range I heard some instructor yelling at his men on the range next to ours. He had a loud voice, and you couldn't miss hearing or seeing him. He was very noticeable with his white mustache and goatee. This was Gunnery Sgt. Lou Diamond. There had been many stories about Lou Diamond over the years. One story was that he dropped a 60-mm mortar down the smokestack of a Japanese destroyer. He was sort of a legend, and quite a character in the marines. No one knew exactly how long Lou Diamond had been in the marines. His sleeves had been saturated with hash marked for years. The Hallmark Hall of Fame had a thirty-minute program saluting Lou Diamond. It starred Ward Bond as the master gunnery sergeant. It came out in the Fifties and was titled *The Man Who Was Two Hundred Years Old*, because stories of Lou Diamond's exploits in the marines can be found dating back two hundred years. I was thinking at the time if I remained in the Marine Corps I might become a master gunnery sergeant some day. This seemed to be the most active and colorful rank in the marines.

After four grueling weeks on Parris Island, the men were split up. Many were transferred to other marine units. I was one of the men who was selected to resume gunnery training in Quantico, Virginia. In Quantico I began coaching marines of various ranks who were preparing to fire in the upcoming division matches. These men were expert marksmen, and I learned more from coaching them than I did from any other type of gunnery training. We became familiar with every type of small arm weapons that were available in the Forties. This included firing all automatic weapons on the FBI range, and learning how to field-strip and clean the weapons. Any time we were not training, we were kept busy grooming the rifle ranges, policing the grounds, pulling weeds, cutting the grass, and pasting hundreds of paper targets on the wooden frames. Over a period of time we had the rifle range looking like a golf course.

I made a number of liberties in nearby Washington, D.C., because the town of Quantico was always crowded with marines.

I met and dated a woman marine while in Washington. We visited the Smithsonian Institute and later took in a movie. We caught a bus back to the women marines' barracks, and arrived just in time for the evening meal. I went through the same mess line, which was quite an experience for me. I was the only male in the mess hall. My date and I spent some time on the base cutting up and having a good time. We lost all track of time, but were alerted when taps sounded. I walked her back to her barrack, passing other barracks along the way. There were no curtains or blinds on the barracks' windows, and most of the women were clearly visible in a state of undress to say the least. As soon as my date went inside, I headed for the main gate. A jeep pulled up beside me, and two women MPs told me to get in the back. They took me to the main gate, and told me that since I had been caught on the base after taps, I would be restricted from ever setting foot in Henderson Hall again. If I was ever caught again I would face disciplinary action. Henderson Hall was located out by the Arlington Cemetery, and I spent most of the night walking back to town along a busy freeway. I never dated another woman marine while in the corps.

On a moment's notice we packed up and boarded trucks that took us to Annapolis, Maryland. We moved into barracks that housed the North Severn marine guard detachment. It was here we were given the reason why we had been subjected to such an extensive training program over the past few months. Our objective would be to instruct the Naval Academy midshipmen in basic gunnery. This proved to be a good and interesting duty. We taught the midshipmen rifle, pistol, automatic weapons, skeet, and trapshooting. We had to be polite to the midshipmen at all times, and we had to call them mister. We were never allowed to swear in front of them. I didn't swear, and I think this was one of the reasons the other men and I were selected for this assignment. The midshipmen were always loaded down with studies and mandatory sports, but most of them were curious and eager to learn about the various weaponry. Calling them mister took away any familiarity. This was good policy because it would have been impossible to remember all the names of the men in the short period of time that we had the many classes. They were well-behaved gentlemen. No one could help liking them. Their enthusiasm knew no bounds, and they always kept you on your toes.

During the two years I spent in Annapolis, I generally made local liberties, but I made two exceptions by making a couple of liberties in Baltimore for two special occasions. These were to see, and dance to, the bands of Johnny Long and Vincent Lopez who were featured in a large dance arena, similar to the Palladium in Hollywood.

No matter what our specialized duties, we still had to pull our turn on guard duty. When on the main gate, this involved sending sailors and marines back to the ship or barracks for being out of uniform. or some other small infraction of regulations. Any time the officer of the day caught anyone in town out of uniform the responsibility for this neglect fell back on the marine guard, so everyone who stood guard was extra careful to observe and check the men as they passed through the gate coming and going on liberty. If any of the guard made liberty in the town of Annapolis, it would only be a matter of time before some of the sailors or marines would confront you. When this happened you had to face up to them the first time it happened, if not your effectiveness as a guard was compromised from that point on. Most of the time I only made liberty on Friday and Saturday evenings, and only then when I could afford an evening out. My regular routine was to take the boat from North Severn, get off at the docks at the naval academy, then walk into town. I would start at the end of Main Street, go through the small town, stopping at certain bars and cafes. I would drink a beer or a cup of coffee and visit with the patrons. This was my way of keeping up with local events, or anything that went on in the town worth knowing, like fights, romances, gossip, and the hiring or firing of employees in various businesses. I would make a number of stops at different places, ending up at the Moose lodge at about twelve midnight. About two in the morning, I would go back to the end of Main Street, enter a small delicatessen for a sandwich and a cup of coffee before boarding the last boat leaving for North Severn. As long as I was single, I very seldom varied from this routine.

Storm clouds were building. It was only a matter of time before marines and sailors would descend on me. I could just about sense the belligerence any time I entered a bar or restaurant. The first trouble that confronted me was in Wally's bar and dance hall. This was a popular hangout for sailors and marines. A big master-at-arms in charge of the North Severn mess hall disliked me intensely and decided to take a swing at me. He missed, and I ended up getting the better of him before the MPs arrived. The bartender told the MPs that the master-at-arms had been trying to start trouble all evening, and no one would fight him. This saved me from being arrested. They took the big master-at-arms with them. I never heard if he received any punishment. Possibly they took him back to the base. The next morning when I took my guard through the chow line, the big master-at-arms was sporting a big shiner. I was able to get my guard through the chow line early, without one word from the master-at-arms.

Another disagreement happened between me and a navy chief over a woman. The chief wanted to talk it over in the head, or rest room, of the Moose lodge. We ended up in quite a battle. The chief had a bloody shirt when he came out of the head. There was no way to hide this incident, and it caused me to get kicked out of the lodge. This chief had been in charge of the Annapolis shore patrol, and the word went around that he was going to get back at me while he was on shore patrol duty. The chief was six feet tall and weighed as much as I did, but once he put a duty belt on, he would have me at a disadvantage, because for me to resist in any way would mean certain arrest and punishment. I had a talk with my commanding officer, Maj. Coggins, and explained what had taken place at the Moose lodge, and the rumor about the chief getting back at me while on duty. The major left to talk to the chief's commanding officer on the USS *Block Island*. When the major returned he told me the chief's face was in such bad shape that he had been granted thirty days' leave to recuperate. The major went on to say the chief had been kicked off of shore patrol for disciplinary reasons, and they had no intentions of taking him back. The major added that he could put the town off-limits to the navy chief if I wanted him to. I told the major that this wouldn't be necessary, because I could handle him, as long as he wasn't wearing a duty belt, and carrying a night stick. The major smiled and said, "Very well, Sgt. Thomas, carry on." I continued to make liberties in Annapolis as usual, but never saw this navy chief again.

The duties of the marine guard were like being in isolation. Many of the non-guard servicemen treated the guardsmen like lepers, and avoided close contact with them when possible. One thing was certain: You could never win many friends while filling the duties of sergeant of the guard. This had nothing to do with the trouble that took place in the Hall of Fame on the Baltimore Highway. This was an after-hours place. When everything closed down in the town of Annapolis, everyone would flock to the Hall of Fame. I had had some strong words for some men who were mouthing off at my date. Sailors and marines who knew me only had instigated and elevated this argument to a higher level. I thought the trouble was smoothed over, but when I left the place, three civilians jumped me in the parking lot. I was able to put two down, and had the third one backing off, when I heard a noise behind me. I turned just in time to see Sgt. Monroe stop a man coming in on my back with a knife. Monroe turned the man around and laid him out cold. There were sailors and marines all over the place, but Sgt. Monroe was the only one who stepped in and gave me a hand. The others were probably hoping I would lose the fight. Sgt. Monroe probably saved my life that night at the expense of receiving a bad cut on his arm. We had to rush him to the Annapolis Naval Hospital for emergency treatment. Any time there was a fight the word spread around town like a wild fire. As a result of a few conflicts I had much less trouble with anyone else.

I continued teaching the midshipmen all that summer, while still taking my turn on guard duty. All the men instructing the midshipmen were to be transferred out in the fall. Only twelve men would be selected to teach the midshipmen during the fall and winter season. I was one of the twelve selected to remain. Some of the other men were able to get transferred into the guard detachment.

The fall and winter months were no picnic. We taught the midshipmen on the indoor ranges, while maintaining the various firing ranges, grooming them to look like golf courses. It took a lot of maintenance with only a few men to do the work. Once a week all the weapons in the armory had to be cleaned an oiled. This was quite a chore, and one I wasn't crazy about, but it gave me something to do inside during the cold winter storms.

I liked the town of Annapolis because it was always alive and moving. It was never stagnant or boring. I have given a few rough and tumble episodes of this tour of duty, but there were a number of memorable experiences that I will always cherish.

Nick, the Greek, owned a hotel and a restaurant on the main street and somehow he became a good friend of mine. He knew Annapolis probably better than anyone else in town. Nick was a good talker, and I, a good listener. He was a big portly man with a good sense of humor. Any time I was in town I would stop for a visit. He would never let me leave without breaking out a bottle of wine for a farewell toast for good luck. Nick invited me to a large Christmas party at his restaurant which was mostly for his close friends and relatives. I considered this quite an honor, since I was the only serviceman attending the party as a non-relative.

The Marine Corps birthday ball was held in Carvel Hall. Cpl. Westerdahl and I decked ourselves out in our dress blues with all our ribbons and prepared to join this annual event. We arrived in town much too early for the ball, so we decided to walk around, to visit, and to kill some time.

A beautiful blonde and a stunning brunette stopped us on the street and introduced themselves. The pretty blonde had recently been crowned Miss Severn of 1947. They were trying to round up two men from each branch of the service to participate in a fashion show at the Annapolis Guard armory. The servicemen chosen would parade around the stage with the fashion models. I joked and kidded with the blonde. I agreed to do it if she would be my date for the Marine Corps ball after the fashion show was over. She laughingly agreed and moved off to locate some more servicemen after telling us what time we should be at the armory. The armory was filled to capacity when we arrived. Westerdahl and I, with the other servicemen, were lined up on one side of the stage, and the models were lined up on the other side. I was the tallest, so I led the parade, walking out and taking the tall brunette model by the arm, parading around the stage. Miss Severn was not one of the models. I had not seen her since I had arrived, so I assumed she had something else more important to do. I asked this tall brunette to accompany me to the Marine Corps ball and she agreed, asking me to meet her when she came out of her dressing room. Westerdahl had also made a date with his model for the ball. On the way out the front door, Miss Severn spoke to me. I had never been caught in a position like this before, so I told her the truth. I had made a mistake and ended up with two dates. I told her that I was willing to escort the two of them to the Marine Corps ball. Miss Severn was very tactful and declined the offer. I felt bad about this, but I didn't know what else to do. I never saw the tall brunette again. I did try and locate Miss Severn to make amends, but I was never able to locate her. I've always hated myself for misjudging someone so pretty. It was my loss.

After the winter months were over, I was assigned as provost sergeant of the North Severn marine brig. This was a very responsible position, and

I had no knowledge of the mechanics of operating a brig. There was a big book on the desk that read, *Rules and Regulations for Naval Places of Confinement*. At the end of two weeks, and after rereading certain passages over and over, I had a fair understanding of what was expected of a provost sergeant.

We had moved out of the marine guard detachment quarters in North Severn, and were billeted on board the USS *Rena Mercedes*, a ship that had been sunk at one time. It had been salvaged, and was now being used as quarters for the marines.

The brig guard was divided into two twenty-four-hour watches. I remained on duty at the brig with my guards and chasers for twenty-four hours and then got relieved by the other sergeant and his watch. They would come across the bay by boat from the USS *Rena Mercedes*. Once my watch was relieved, we caught the waiting boat for the trip back to the ship. We alternated on the weekends. Each watch remained on duty for the entire forty-eight hours, and then had the next weekend off. This was the general routine we followed.

That was one position I could never relax in because I was always busy. Prisoners were being worked all over the base. I not only had my guard to worry about, but also the men they were supervising.

When I took over the brig the men didn't have any linen, cigarettes, or toilet articles. I set about to change this. These were not hardened criminals, but servicemen who had overstayed their leave, or made some other minor infraction of the regulations and rules. The most extreme punishment seemed to be thirty days on bread and water with a full days' ration every third day. I called the lieutenant who was the provost marshal a number of times to try and get some assistance. He never came over or gave me any advice. He always let me know he was busy, and I was to handle anything that came up. I believe his attitude, more than anything else, contributed to the trouble that eventually took over the brig.

The guard was not to show favoritism to any one man in the brig, so the only way I could figure out how to get them cigarettes was to have one man buy one carton of cigarettes and make those available to all. After they were gone, I would let another prisoner buy a carton and start the same distribution all over again. This was the best I could do, not having any support from the provost marshal's office.

About all the prisoners had for linen was a dirty blanket. I started requisitioning more than the necessary linen for the guard, and using the surplus for the prisoners. The Red Cross refused to help me in any way. They could not under these conditions assist able-bodied men, or so they said.

About a week after I took over the brig I began to sense that something wasn't right. I was judging from the actions and mannerisms of the prisoners. It started with small things—a bruised eye, or a bleeding mouth. Sometimes a man wouldn't eat, or said he didn't want to go to chow. Any time I inquired about these oddities, the man in question always had an excuse, like running into the brig door.

The brig office was located across the street from the brig. I shared a large desk with the sergeant on the other watch. Both of us kept our side of the desk locked. I started a file of notes on every irregularity that came to my attention. I always included a copy of the brig guard, date, and roster of the prisoners in the brig at that time. I continued this as an ongoing thing, all the while trying to find out what the problem was, and how to solve or stop this abuse. I called the provost marshal a number of times after sensing this abuse to get him to come over and talk to the prisoners. He continued to say that was my duty, and I should be able to take care of these things.

I called all of the prisoners together and told them I suspected some of them were being mistreated. I told them that if anyone ever struck or hit me while I was confined to a brig, and unable to fight back, I would do my best to kill them if and when I ever got out. I meant this, and I think the men knew it. I went on to explain that if any man came forward with a complaint, I would bypass every rank and take them directly to see Admiral Holloway, the director of the naval academy. I had the authority to do this, but no one came forward. I tried talking to individuals in private, but I still could not glean any information from the men. I told the gunnery sergeant at the armory of the suspected abuse and asked him to come over and talk to the men. Since he had no connection with the brig, I thought perhaps it might be possible that he could get someone to talk. He made a nice appeal, but we still came up empty-handed. I tried something else. Every time I detected something wrong and wasn't satisfied with the answer, I would take all the prisoners out and give them close order drill for four hours without stopping. I tried to wear them down, but after a week I gave this up, because I was also doing close order drill without a break for the long four hours.

One evening when I had the prisoners fall out for the evening mess, I found the newest prisoner bleeding from the mouth. I took him to sick bay, and later to the naval hospital. I interrogated him all the way back to the base, but he refused to tell me anything, except that he had run into the brig door. This ceased to be an excuse, because any time I had the prisoners fall out for anything, I personally made sure the door was fully open and locked back.

All the mail leaving the brig had to be censored. Most of the time it was routine: a letter to the wife, or to the family. There was seldom anything to cut or blot out. A number of times the prisoners would receive letters addressed to the USS *Marine Brig*. I guessed the prisoners had told their wives or sweethearts that they were doing duty on board a naval vessel. One morning I was censoring the mail as usual when I came across a letter from the prisoner I had taken to the hospital. It was addressed to his mother. It contained all the information I had been seeking for so long. It told of a kangaroo court in the brig, condoned by the sentries and guards. The man described his ordeal of having his hands tied to the overhead pipes in the head, and then being beaten in the stomach. This was all the information I needed to go to the admiral. Before leaving, I decided to check the manual on censoring prisoners' mail. According to the rules and regulations for places of naval confinement, mail is privileged information—you either censor the mail, or let it go through. You are not at liberty to divulge any part of a personal letter to anyone.

I hated to let this letter get out of my hands, but it was addressed to his mother, so I decided any action taken would have to be initiated by his mother. I mailed the letter without censoring one word. I made a copy of the letter with the man's name and his mother's name and address and the date it was mailed. This was filed in my desk drawer with all my other notes and infractions.

One day I got a call from the provost marshal. He said there was a prisoner on the way over, and I was to make it rough on the man while he was in the brig. I signed the release papers for the prisoners and took him inside to package his personal belongings. I told him about the phone call, and I asked him why it was made. He said he had made a lieutenant out to be a liar at his court martial. I told the man that as long as he fell in with the others in following the brig routine, I wouldn't treat him any differently from the others. He thanked me, and he never gave me any problems while he was in the brig serving his time.

I had prison chasers working prisoners all over the base: the galley, the mess hall, the grounds, the bowling alleys, and the recreation rooms. I would frequently make surprise inspections, not only to catch the chasers trying to do something out of the ordinary, but to keep them on their toes. I thought I had some good men, and I couldn't imagine any of them mistreating prisoners, but someone was doing it, and I was hoping I would have the answer very soon. No mother in the world would put that letter away without notifying someone. I thought perhaps it would be sent to the admiral of the naval academy. It had been almost three weeks since I had mailed the letter, and I was beginning to think nothing would ever become of it.

It was about nine o'clock in the morning. I had just completed the daily brig roster and had picked up the newspaper when in walked a full admiral, two marine generals, and all of their staff. I almost fell out of my chair. I stood at attention. The admiral picked up my roster and asked if this was an up-to-date list of the men in the brig. I answered in the affirmative, and the admiral sat down and said, "I want to talk to these men; starting alphabetically, bring me the first man on the list." He continued to talk to these men behind closed doors. After he had talked to about ten prisoners, he called me into the office. The admiral said, "Sergeant, you have a prison chaser and a sentry on your watch," and he gave me their names. He ordered me to lock them up in solitary confinement they were awaiting general court martial. The admiral also named two prisoners in the brig. I was to lock them up in solitary. They, too, would be awaiting general court martial. If anyone tried to talk or communicate with these men in any way, I was to arrest them and put them in solitary confinement. They, too, would be awaiting general court martial.

All of the high-ranking officers departed in their limousines. This was the first time an officer had visited the brig since I had been the provost sergeant. I thought the officers were going back to D.C., but one hour later a boat came over from the USS *Rena Mercedes* with the provost sergeant from the other watch and three of his men. I put them in solitary confinement; they were awaiting general court martials. This was really a big surprise to have to lock up a sergeant from the other watch. The sergeant was a good friend of mine, and I never thought for a minute that he would be involved in brig troubles. I learned later that he had received the same phone message as I had to make it rough on a prisoner. As I understood the charges, he never touched the man but had prisoners in the brig beat the man.

If the investigating admiral had asked me anything about the administration of the brig, I would have given him the file with all the information that I had been collecting. If he had opened the drawer to the desk, which was unlocked, he would have discovered my notes. The officers departed without asking me one question about the brig.

The next day the whole world knew about the naval academy brig troubles. All of the papers including the *Washington Post*, had headlines that read "Marine Sentries Beat Prisoners." The investigation and the resulting court martials were the top headlines for over a year.

The investigation continued, and some of my men told me that the investigating team asked many questions about me, and the way I ran the brig. I was worried that the two men on my watch, who were facing charges, might be tempted to say they were just following my orders.

One morning I passed the provost marshal while on board the USS *Rena Mercedes*. He stopped me and said, "By the way, Sgt. Thomas, on this brig investigation, you are the only one whose name is completely clear of the whole thing." I told him that I was glad to hear this. He said, "I wish I could say the same thing." This officer ended up being transferred out of Annapolis as a result of these investigations.

When the court martials convened, I was called in and asked to testify to something that wasn't exactly true. I told the officers who were acting as defense attorneys that I wouldn't, and couldn't, do it. They told me I would be charged with direct disobedience to a lawful order if I refused to testify the way they wanted. I told them to take whatever action they cared to, but I wouldn't testify the way they wanted me to. They became very irate, and yelled that if this was the way I was going to be, they didn't want me as a witness, and to get the hell out of there. I didn't expect to be treated like dirt just because I wouldn't bend to their wishes. I left with a burning rage swelling up inside. I wished I could fight back. The retaliation I had in mind would have been devastating.

I was still shaking with a burning hatred when I left the ship. I caught the next boat over to the North Severn brig office and removed all my notes and the file from the desk drawer. I then went around behind the mess hall, tore them into unreadable pieces, and threw them in the trash bin. Later I was asked to make the same fabricated testimony for two other prisoners, but I still refused to do it. The defense was trying to lay all the blame on the higher officers. The officers were in no way responsible for the men's misconduct. The men knew very well what they were doing, and I believed they should be the ones held responsible for their actions. Most of my notes may not have been admissible as evidence, but I had the names of the men who were in the brig, or present at certain times, who could have corroborated many events and observations. Since I would not testify to anything but the truth, I was never called as a witness for any of the court martials. The sergeant was acquitted after a lengthy court martial. Of the two men on my watch who stood a general court martial, one was acquitted and the other man received a two-year sentence and a dishonorable discharge from the service.

After the brig scandal, all of the men who had been connected with the brig were transferred to some other duty. I was transferred into the guard company. I had already been standing guard, and since I was familiar with all the posts and duties I had no trouble adjusting to the new duties. Instead of taking my turn on guard duty as before, I was now on duty as sergeant of the guard most of the time.

I'll not list every small encounter that most often seems to develop as a result of a guard enforcing his duties and carrying out the many special orders from the officer of the day or the base commander. As time elapses, and you have a general turnover in personnel, past conflicts seem to fade out. When this happens it's very easy to have a repeat of the same old problems. I'm a firm believer in getting my guard on post at the proper time; this entails running my men through the chow line as early as possible. The mess hall was run entirely by the navy and always had a master-at-arms present controlling the chow line. I don't know where the navy got the master-at-arms, but they frequently seemed to be arrogant and often took pride in aggravating me as much as possible, just because they knew I was in a hurry to get my men on post. At times the master-at-arms had me wake up both the officer of the day and the commanding officer to get a chow pass authorizing me to take my men through the mess line—anything to delay me. This was a repeat of the problems I had with another master-at-arms. This master-at-arms was fairly new on the job and didn't know anything about the previous run-in. One evening while on liberty, I had a chance meeting with him while I was leaving a rest room. He must have thought he was still on the base when he made a nasty remark to me. I knocked him flat on his back. He was a big man, and when he got to his feet, we went round and round until the manager broke us up. The following morning I had my guard lined up as usual at the crack of dawn waiting to go through the chow line. I was sporting a bruised eye. The master-at-arms also had a shiner, but his was more pronounced than mine. A small victory, but the master-at-arms never gave me any more static while he was in charge of the mess hall. Word gets around, and my chances of having any more trouble were greatly reduced. I didn't have the most popular post in the Marine Corps, but the men who disliked me because of my duties seemed to respect me as long as I treated everyone alike.

During the time I was in Annapolis, I stood honor guard for President Truman and other admirals and VIPs. Because I was tall I was usually at the head of the rank. I had my picture taken many times when dignitaries inspected the ranks. I never knew what happened to those pictures; I never got to see any of them.

The first honor guard I stood for President Truman was the roughest. The first sergeant had us on the docks at five in the morning, in freezing weather with a strong wind whipping across the bay. We were decked out in our light dress blues. We stood on the frosty dock until we were almost numb, and then did close-order drill to thaw out. This was repeated until the president's yacht docked at about ten o'clock. The next time I stood honor guard I was ready. I had wool socks and a heavy set of long underwear.

One evening the duty NCO came into the barracks looking for the navy man who had been baby-sitting for Comdr. Grey. A short time later, unable to find the man, he returned to the barracks and asked if anyone was interested in a baby-sitting job for Comdr. Grey. Having nothing to do for the evening, I volunteered for the job. I was used to caring for many smaller brothers and sisters at home, so I felt qualified for the job. I was surprised when a long limousine arrived at North Severn to take me to the commander's quarters. This was a big two-story building inside the naval academy. Comdr. Grey was the admiral's aide, a very distinguished officer and gentleman. He had a very pretty wife who was very gracious and offered to cook me something for supper, even though she was already dressed for a cocktail party. The commander was quite worried about fire because of the two-story frame building. He wanted me to put his two small girls to bed at exactly eight o'clock and make periodic checks on them during the night. The two small girls were pre-kindergarten darlings and very well behaved. There was a small TV in the house, which was still a novelty. It fascinated me at the time, and I enjoyed watching the shows. It kept me awake, and I checked the girls every thirty minutes. I don't recall how many times I baby-sat for the commander, but I enjoyed the job and just being away from the barracks for awhile. There were other marines who wanted to take over my job. They watched in awe every time I departed in the long limousine.

One morning the first sergeant called me into the office and told me to get dressed in my dress blues and to report to the admiral's office. I was to be interviewed for the job as the admiral's orderly. This was a coveted position for which most marines would give their right arm. You received an allowance to live off the base, plus many other special privileges. I told the first sergeant that I didn't want the job as orderly. He said I had to go since the admiral's office had requested me. Another man who was interested in the job was told to get dressed and go with me. I was interviewed by Comdr. Burke, who thought there was something wrong with me not wanting the position. The other marine was interviewed and was accepted for the position. A few months later he was recommended for an appointment to the academy. Sad to say, he never completed the training. He made the football team, and in one game he had sustained an injury that caused him to be dropped from the academy. I never heard what happened to him once he left the school.

While in Annapolis I met, dated, and married a pretty redhead from Fort Wayne, Indiana. Being a married marine, I was allowed to live off the base. About this time the Marine Corps had a cut in appropriations and had begun cutting the Marine Corps personnel. All married men would be able to take an early discharge if they wanted. I had run out of wars to fight, and the marine guard had become boring. Being a married man, I decided to call it quits and try to succeed in a world of civilians. It was time to settle down and get a regular job. This would be the only way I could ever have a stable married life. I took my discharge and joined the inactive reserve. With no war to fight, there was no reason to think I would ever see combat duty again.

The employment office in Dallas hadn't changed one iota. They were still an equal opportunity employer. They treated everyone, including veterans, like dirt. They refused to let me put in for unemployment as long as there was a job that I qualified for. Since I didn't have a trade, I was qualified for every unskilled job in the state of Texas. I considered the employment office to be an enemy, and I stayed clear of them for the rest of my working career. I wasted some time on part-time jobs before deciding to attend school and really learn a trade.

My marriage lasted for three years before it ended on a sad note. After my freedom, I entered the Institute of Radio Broadcasting and Engineering. To supplement my school subsistence, I got a part-time job working for the *Dallas Times Herald* driving a truck in the evenings. I decided to work and stick with my studies, ignoring all other types of activities. This would be the only way I could ever get ahead in civilian life. If I was going to succeed in anything, I would have to settle down and apply myself to the present situation. It would entail practicing self-discipline and denying myself many small luxuries.

I'm going to give the reader a brief glimpse into my educational background before continuing with my narrative. This explanation will help you to understand some of the decisions I made and would continue to make.

I started first grade at seven years of age in an old country school in Strip, Texas. This was a small, two-room frame building that had the first to fifth grade in one room, and the sixth grade in the other room. We had a pretty young woman as our teacher. The sixth grade was taught by a man, who was also the principal. Discipline was very strict; if you did something wrong you received a whipping. The whole school had to witness a flogging of one of the sixth graders. He was whipped with a strap until the back of his shirt and pants were covered with blood. I never learned why he received the whipping.

This close association of the different grades had some advantages for a first grader. If you paid attention and had the ability, you could learn just as much in the first grade as the fifth graders. I started reading fourth- and fifth-grade books while I was still in the second grade. I always thought schools were much too slow, wasting too much time on one textbook. I always read ahead, no matter what the subject matter happened to be.

When I reached the third grade, my parents moved to Dallas. I was a country kid, and must have looked awful on my first day of school. I remember the teacher telling me not to touch the walls because I would get them dirty. I finished the fourth grade and was then double-promoted to the sixth. I completed the sixth grade in the Rodger Q. Mills Grade School in Trinity Heights and entered the seventh grade at James Bowie Junior High School in Oakcliff. After completing the seventh grade, I had to drop out of school. We had a house full of kids, and my dad just couldn't afford to send my brother and me to high school. I was still hoping there was some way I could start school when it opened, but my dad didn't have any money for clothes or school supplies. It was bad enough that my brother and I couldn't attend school, but we had to put up with a neighbor who would come over and continually brag about how her two boys were doing in sports and other school activities. It was like a sore spot to me. Every time the neighbor came to the house, I would make a quick departure. I wasn't interested in anything that person had to say.

I read a lot, trying to advance my learning with the hope that I would be able to enroll in school the following year, but this was always out of my reach in our present predicament. It wasn't long before I started getting into trouble with the juvenile authorities of Dallas, and I eventually left Dallas for good.

Here I was in Dallas again, but this time I was trying to advance myself by attending the Institute of Radio Broadcasting and Engineering. I entered the school with enthusiasm, striving to be the best in the class. I had only been in school for about six months when I started to MC live talent shows and host two record programs a week on local radio stations. I had to do all the continuity writing and the commercials for the live weekend talent shows and for the two weekly record programs.

My programs did so well that I started reserving a table every Saturday night at the Roundup Club on South Ervay. I would let the club seat the guest stars at my table, and when I could get away, I would go to the club and interview whoever the country celebrity happened to be at the time. The country artist would generally send me some records that I used on my programs. If I had a live talent program, I would have the Sellers recording studios in downtown Dallas make recordings of the program an put them on 12-inch discs so I could play them back.

Everyone on the programs was taken to the Town and Country Supper Club by potential sponsors for a big steak dinner. We reached an agreement for a two-year contract for a one-hour weekly live talent program. I would be the continuity writer and announcer for the shows. The deal was approved by the station, and we were all set to sign once the contract was completed.

I had an old Studebaker that was paid for. I took it down and traded it in on a newer car. With new shows on the horizons I would be able to afford one. I was proud of my new car, and decided to drive across town, see the folks, and show it off. I had been very busy, and I didn't get over to see my folks very often. While I was there for the visit, my mother gave me some mail that had been there for quite some time. It was orders to report for active duty for Korea. This was hard to believe. The inactive reserve was being called up before the active; it didn't make any sense to me. I only had three days to make it to Camp Pendleton, California. I parked the car at the boardinghouse, since I would not be allowed to take it out of the state. I borrowed money from a girl in the boardinghourse, leaving two tall stacks of records and recordings as collateral. I took some bare essentials and hopped a Greyhound bus for the Far West. Later when I sent someone over to turn my car into the finance company and reclaim my records, I learned the girl had married and moved to Arkansas, leaving no forwarding address and taking my records with her.

I had time to reflect on my civilian efforts in Dallas. I had completed the school of radio broadcasting and was well into my studies in engineering. I had interviewed a number of celebrities in country music and many dignitaries in state and local government. I had spent over a year training my voice to eliminate most of my Texas drawl. All of this would be a waste of time. Once I started yelling at troops again, I would fall back into my old Texas drawl. Once again my glorious civilian life had come to an abrupt halt. All my hard work and training for two years had just been wiped out.

Camp Pendleton and the Pacific Crossing

Camp Pendleton still looked the same and hadn't changed. We were processed and went through classification. I requested a job in public relations, but the classification officer said, "We have enough public relation marines walking around doing nothing, so I'm not going to change your artillery specification number."

After processing we were assigned to Tent Camp Two, a combat training camp located in San Clemente. I was out of shape and needed this training really badly before joining some combat unit. This was strenuous training, with many long hikes up and down mountains. Once again I started utilizing one canteen of water for every twenty-mile hike. I had to take advantage of this training and try to reclaim some flabby muscles. I still believed in practicing self-discipline. There might come another time when I would have to endure with only one canteen of water over a long period of time, like I did in the Pacific. Each time we came back from a twenty-mile hike, and the men were dismissed, they would knock you over to get to the cold water fountain. I always had water in my canteen, and I never drank cold water if I was hot and tired. At times I got some cold water and sat and waited until it warmed up to room temperature before I took a drink. I couldn't say even now if I benefitted from this practice, but I don't believe I lost anything.

Tent Camp Two was far from a picnic. It was like another boot camp without the buckets and the sand. All of the instructors tried their best to intimidate trainees. These instructors taught a number one course. They were combat veterans who knew what to teach and how to teach it. When they yelled to fall out, they expected us to fall out faster than the boots in boot camp. At one roll call, one marine went out so fast he ricocheted off the door and stepped onto one of the first sergeant's flowers. That evening after chow we were given the choice of revealing the man's name or suffering the consequences. When no one volunteered the name of the man, the first sergeant had us fall out at six that evening, and we were given close-order drill for six hours without stopping. Every unit in our camp screamed for the sergeant to knock it off so they could get some sleep.

Most of the marines in this camp had nothing to worry about except their training. I though this was all I had to concentrate on, but one night while on guard duty I got a big surprise. The corporal of the guard drove out to relieve me on post with the supernumerary of the guard. The corporal gave me a phone number in San Diego and told me to call my wife; it was an emergency. I was sure they had the wrong man since I wasn't married. It was about midnight when the corporal of the guard dropped me off at the only phone booth in camp. Even at this time of night, marines were lined up to call home. I got in line, waited my turn, and then called the number in San Diego. It was a girl I had been dating at the boardinghouse in Dallas. She had caught a train to San Diego and had spent all of her money on a hotel room. The rent was due before checkout time, which was eight o'clock in the morning. I only had a few dollars and told her I would see what I could do and call her back. I must have had a puzzled expression on my face when I came out of the phone booth, because a marine waiting in line asked me if I had trouble. I told him my problem, and he said he could loan me twenty dollars. He told me his name was Tankersley, and he knew me from the pay line. I accepted the loan and called the girl back. She had enough money for the bus fare to Oceanside. I told her I would meet her there. I got a couple of rides, but walked most of the night. Once in Oceanside I rented a room and then met her at the bus station.

The next morning I went to the Red Cross to apply for a loan. All they would loan me was ten dollars. The following payday, the pay officer told me an official from the Red Cross was waiting across the street to collect the money I owed them. This was embarrassing, but I paid Tankersley and then went across the street and repaid the ten dollars. I couldn't believe the Red Cross would send a man thirty-five miles one way to collect a ten-dollar loan.

My girlfriend and I ended up getting married just prior to my departure overseas. We had a very good relationship while living in Oceanside, even though much of my time was spent in training. My wife of a short duration left for her home in San Antonio just before my training was completed. I kept a clipping for years that read, "Bridegroom off to Korea."

Our training unit moved from Tent Camp Two in San Clemente into a barrack located in the back part of Camp Pendleton. The first thing I noticed about this location was another and much shorter road leading into Oceanside. It would have been nice if someone had told me about this road years ago. I had only known one road in and out of Pendleton, and I had hiked many, miles on this road to go on liberty.

We spent the better part of one day drawing our cold weather gear and getting the correct stencils on our sea bags. Our regular sea bag would be put in storage in Japan until our return from Korea. The cold weather bag would accompany us on the ship. We were told that the cold weather gear cost the government over seven hundred dollars per man, and if we lost or misplaced this cold weather clothing before we arrived in Korea we would have to pay for it.

We received instructions to stack our sea bags at two o'clock the following morning. One stack was for the cold weather gear, and the other for bags going to storage. We would then board trucks to San Diego for embarkation.

This was my last night in the States. I went into Oceanside for one last time, just to get off the base. I arrived back at the barrack about twelve o'clock. Many of the men were carrying their sea bags out and stacking them. I decided to take mine out before turning in. While I was in Oceanside, someone had stolen my cold weather bag. I was hoping someone had taken it out, and stacked it for me, or maybe it had been done by mistake. It was too late to do anything about it then, so I decided to keep quiet for the present.

I was surprised to see all of the changes that had taken place along the highway leading to San Diego.

Camp Elliot was gone. I had spent a month of rough training in that camp before going to the Pacific in 1943. I recalled the camp had one monster of an obstacle course. Live ammo was fired a few feet above your head while you crawled through the fine red dust. This dust stuck to your sweaty clothing while you crawled and worked your way through and under barbed wire. Explosives detonated all around you, spraying fine dust over everything. When you came out at the end you were as muddy as a pig, with the sticky goo all over you and your equipment. It took many hours of hard work to get your clothes and weapons back in shape before the next inspection. We had one man who had deserted while in that training. That had happened right after we went through the obstacle course. No one knew what happened to him. He was bunked in our small barracks, and we thought he might be a victim of foul play. All of his equipment was accounted for except his rifle. It was a week before the military authorities caught up with him and brought him back to the barrack. He crawled up into the small attic and retrieved the most God-awful rifle I have ever seen in my life. It was hard to recognize this piece of crud as a trusty marine's friend. It had been stored away with sweat and dust all over the weapon. Believe it or not, this was why the man had deserted. He didn't think he could ever get the weapon clean, so he deserted the marines rather than tackle the monstrosity that was supposed to be his weapon. He was taken out of our unit, and I never heard what became of him. Up until that incident he had been a good marine.

Linda Darnell and Keenan Wynn put on a memorable USO show for the troops in Camp Elliott. I remember Linda Darnell was so much prettier in person than on the screen.

The paratroopers were always an impressive sight in Camp Elliott, double-timing everywhere they went. I tried to put in for paratroopers but was told the men were selected from boot camp. My only interest at that time was to bail out in a parachute.

Old Camp Elliott was no more. Thousands of marines had undergone training there, and I m sure many of their memories would be much more interesting than mine.

Once again I was embarking in San Diego to fight another war. I was surprised to see so many people on the docks to see us off so early in the morning. Many were relatives or friends. The marines were all dressed alike, so it was hard for the people on the docks to recognize anyone. I didn't have anyone to see me off, but I remained topside and waved right along with the rest.

Raising anchor and getting underway is all hustle and bustle as all the sailors perform their assigned duties. Very slowly the ship slinks away from the docks and is soon slithering through the bay. Within an hour we would be in open water, bound for our date with destiny with yet another enemy in a different land.

Tankersley, my acquaintance in Tent Camp Two, had procured a small plug-type chess set with instructions. We used this to pass some of our time. The ship had a small library, but it was not open for the troops on board. When not playing chess, I would go up on the bow of a ship, as I had done many times before, and ride the bow up and down as the ship plunged ahead through the restless waves. Flying fish broke the surface, and skimmed in front of the ship just a few inches above the water. It was like observing a race to see which one could remain aloft for the longest period of time.

In recrossing the cold blue Pacific, I had plenty of time to reflect on many events that had taken place while fighting in the islands in World War II. As the ship ploughed ahead, a montage of memories flowed through my mind as we passed north of many old camps and the small islands that had been so very costly in American lives.

New Zealand Remembered

It was well after dark when the truck dropped me off in the Second Marine Division camp. I was assigned to F-2-10. Early the following morning I went outside to look around the camp. It was a very cold and foggy morning. I walked down the company street between two rows of tents, feeling very depressed. I was like a lost soul, not knowing anyone. I had the feeling it was me against the world. A lone marine was coming toward me down the company street. When he got even with me he said, "Welcome aboard, Thomas." This was the gunnery sergeant for F-2-10, the youngest gunnery sergeant in the Marine Corps. The gunny would not let anyone grow a mustache or beard since he was unable to grow one. He would make it a point to learn the names of all the new men so he would be able to call them by name. Anyway, it sure gave my morale a boost. I was feeling much better when I fell out for my first meal in F-2-10. Sgt. John Paul Young was an excellent gunnery sergeant, and we all felt the loss when he was wounded and had to be evacuated during the battle for Saipan.

New Zealand looked like the Garden of Eden with all the lush green grasses covering the beautiful hillsides and mountains. We had many maneuvers while in this paradise, and I always hated to disturb or dig a hole in the silky grass.

We were on one of our infantry field maneuvers against the aggressor forces in a rural area of New Zealand when our platoon came to an old country farmhouse. This house stood between us and the aggressor forces. The house was surrounded by a vegetable garden and a large flower garden. All of this was enclosed inside a fence. The sergeant in charge had a brilliant idea. He would send a squad of men over the fence and through the garden. This would catch the aggressors by surprise. The squad leader had his men over the fence and about halfway across the garden when an elderly woman came storming out of the house, swinging a large broom. She started chasing and swatting the marines, hitting them across their backs and on their helmets. It was very comical to see a bunch of big marines falling all over themselves, trying to dodge the woman and get back over the fence. Needless to say, we lost our chance of a surprise skirmish to the opposing force.

I could never think of New Zealand without remembering a beautiful sweetheart. Not knowing her name, it will always remain a cherished memory.

Life aboard ship remained unchanged. Anyone might think a cruise aboard a troop ship was like a pleasure cruise, but we were kept busy with inspections, guard duty, and abandon-ship drills. Everyone got his turn on guard and mess duty. Mess duty aboard ship is complicated by the constant rolling and pitching of the ship. The heads, compartments, and our bunks were inspected every morning except Sunday. Inspections aboard this ship would be few because the ship was moving at a fast clip. It was doing its best to get us to the war before it was over.

Hawaii Remembered

It seemed like ages since I had left Hawaii, but the many memories were still very vivid in my mind.

The most notable was that of arriving in Hawaii from New Zealand, and noticing the extreme and stark contrast between the two islands. We went from lush valleys, mountains, and farm lands to the barren sands and deserts of a volcanic island.

Many thoughts kept coming back from our stay in Camp Tarawa, some humorous, and some sad and pitiful.

I remember that most of the men in our gun section talked about their girlfriends back home. Girls, school, and sports were usually the topics of conversation. At the time everyone in the gun section had a girl waiting back home except Bill Moss and myself. They were always bragging about the girls. I was a loner, and was generally left out of most of their conversations. It was like I was sitting on the sideline watching a stage presentation unfolding.

The first "dear John" letters started arriving while we were in the camp. When one man would receive a "dear John" letter, the other men would make it worse by teasing and ribbing him unmercifully. Eventually the "dear John" letters made the rounds to every man in the section, with the exception of Leroy Sorenson. Sorenson would torment and tease the other men endlessly, and he would continue bragging about what a sweet and dedicated girl he had waiting for him back home. After one mail call, Sorenson came into the tent reading a letter. Suddenly he swore, threw the letter down, and stormed out. The men in the tent picked up the letter and read it. His sweet girl back home was getting married. All of the men went yelling this up and down the company street. Poor Leroy had gotten a "dear John" letter. This concluded the mail from the girls back home for him. You couldn't help but feel sorry for the men in our gun section. All of them had taken their "dear John" letters very hard.

We had a gunner in our battery by the name of Wray. Gunner Wray wore the bursting bomb insignia on his shoulders, signifying his rank. He had spent most of his life in the Marine Corps, and had the command presence of an English bulldog. A conversation was practically non-existent with this gunner. He spoke mostly in commands and orders, always in a loud voice and with a scowl on his face.

Pulp magazines found their way into our tent, and we would read them and pass them around. One evening one of the men was reading a *Ranch Romance* magazine. When he came to the pen pal column, he read some of the letters out loud. He read one letter from a middle-aged woman who lived alone on a farm; she was crazy about horses, cows, pigs, and chickens. She was looking for a gentleman to live on a farm who also loved horses, cows, pigs, and chickens. I'm not sure who advanced the idea to write to this lady, but we all sat down and composed a nice short letter. The letter read, "My dear lady, I'm a Marine Corps gunner, and I plan on retiring from the Marine Corps as soon as I possibly can. I may have to wait until the war is over, but in the meantime we could correspond and get acquainted. I love farm life, and I just adore horses, cows, pigs, and chickens. If you are interested please respond, as I would like to get married and settle down on a nice farm." We signed the gunner's name and address to the letter, and mailed it. Everyone waited and wondered to see if Gunner Wray would receive an answer. We couldn t find anyone who could muster enough courage to approach the gunner and ask him if he had received an answer. It would have been nice to know the outcome of this. It's possible the gunner has retired, and is now living on a nice farm with a nice lady and many horses, cows, pigs, and chickens.

Some men under extreme stress often suffered combat fatigue. They often displayed deviant behavioral patterns or symptoms. I don't believe anyone can predict behavior. Many of these men were sick and overstepped the realm of reality. They were in a world of their own fantasies. Some of their acts were pathetic, and you couldn't help but feel pity for the men. Two MPs caught one of our men on the outskirts of Kamuela with a horse. They took him to the base hospital. I wasn't there, so I have no way of knowing what really transpired. I'll leave this to the reader's imagination.

The information of the marine's aberrant behavior went out over the division switchboard like greased lightning. It would have been impossible to hush it up. This proved to be a big embarrassment for Fox Battery. I'm not sure when or where it began, but back through the ages the marines have called cold cuts, or processed meats, horse cock. This is only one of the many slang words in the marines' language. After this incident, every time our unit was marched to chow, the other units of the Tenth Marines would start yelling and jeering, "Break out the horse cock, here comes Fox Battery." The din was so loud that the other NCOs would have their respective units remain at attention so you could hear their commands to file into the mess hall. Any member of the Tenth Marines in Hawaii at that time would remember this. It wasn't the easiest thing to live down or forget.

Tarawa Remembered

In reminiscing about the island fighting, it would be hard to forget all the men who died fighting for the small island of Tarawa. Many of the men died wading in after being stranded on the coral reefs that surrounded the island. I was eighteen at the time, and most of the marines who took part in this landing were about my age. It was a pity so many had to die, but I thought it was awful that these men who died wading to shore under the withering fire never had a chance to prove themselves in closing with the enemy. Some of the best laid plans can go awry, but when it happens in an amphibious assault landing, it's paid for dearly with the bodies of dead marines.

Tarawa was a battle that forever changed the lives of the men who fought there. It changed the life of Bernnie, my brother, who was severely wounded on the island. He would never be back to normal. He had over twenty operations on his face and was now wearing a glass eye. He was one year older than I, but I always looked upon him as a younger brother. He never liked to fight, and many times I would have to intervene and take up for him. This started when I entered the first grade. I was exposed to fights and death while in the first grade. It's hard to say if this has been an asset or a liability. Bernnie attended school for almost a year when I started in the first grade. He warned me about a school bully in the third grade by the name of Dock McCracken. He wore overalls, had a freckled face, and really looked mean. He had beaten up and chased just about everyone in the school at one time or another. After my first day at school, he got after me and chased me for a long way before giving up. At the supper table that evening I told Bernnie I didn't know why we ran from him, because we were as big as he was. My dad, overhearing this, said he had better not catch us fighting on the school grounds, or we would get a hard whipping when we got home. I decided to take Dad at his word. When school was out the next evening, I stuck my tongue out at Dock and took off running down the single-lane road. Well away from the school grounds, I turned and crossed a cattle guard, ran out into the pasture, and stopped. I turned around and gave Dock about three hard punches to the face just as he caught up with me. His nose started bleeding; the fight was over. He never bothered any of us kids again as long as we attended that school.

I lived two miles from school and developed a close friendship with a German boy by the name of Owen. He lived about halfway between the school and my house. We were almost always together. Going home from school in the evening, we would always be on the lookout for a dust cloud that rose up behind the only car in the county. This would be the teacher's brother coming to take her home. He always passed at a high rate of speed, throwing dirt and dust all over us as he swooped by. One evening, like all the rest, I saw the dust cloud approaching. I moved off to the side of the road. The car would have hit me if I hadn't jumped into the ditch by the roadside. I heard a loud thud, and school papers flew all over the place. Other kids along the road took off running. The teacher's brother picked Owen up, put his limp body in the trunk, and slammed it shut. He saw me watching and said, "Get the hell out of here." He got in the car and drove off. I attended my first funeral for my first friend. Nothing was ever done to the teacher's brother. A few months after this, I attended the funeral of my sister who was killed in another accident. I had three sisters, but I had been very close to this one.

I have always been sorry I wasn't with Bernnie when he was wounded on Tarawa. All the hospital ships were full, and he had to remain on the island with severe wounds for almost twenty-four hours before they were able to evacuate him to a hospital ship. The pain must have been unbearable. During this wait he was blinded in both eyes from powder burns.

Saipan Remembered

When we were in the vicinity of the Marianas, I remembered all the dead marines that we buried in the division cemetery. It was very neat, with all the markers in neat rows. The tropical growth of the island is so rapid, it would be possible the graves were grown over by now. I hoped someone on the island was keeping them up. I had spent over a year on and off of this island. It was on this island during combat, and after picking up some of our dead and wounded, that I discovered something that was impossible for me to do. I could not smile, no matter how hard I tried.

I suppose at the time everyone in Fox Battery like myself was anticipating this assault landing in their minds. It was hard to think of anything else. My nerves were stretched to the breaking point. This tension continued to build all the way to the beach. You tried to prepare your mind, and your way of thinking, to conform to the perils that you know lie just ahead. At the same time you develop a mental will, or a sense of survival, that will sustain and carry you through the most traumatic and confounding circumstances that you know will confront you once you have landed. No matter how hard you have tried to prepare and condition yourself on surviving, there can be no guarantees. A sniper's bullet, or a piece of shrapnel, could shatter everything in one second. You have learned what you could, and hope this knowledge will be enough to carry you through any perilous ordeal. This and a million other thoughts are coursing through your brain as you near the beach. All of these built-up emotions, suspense, and fears will be dispelled immediately once your feet hit solid ground. No more imaginary fears; you are committed to the battle. It's either fight or be killed; there is no other alternative. Doing your assigned duties and trying to survive has priority over all other thoughts. You follow through, doing the best you can with whatever you have

103

available at your disposal.

This war was already fought and won. I had been one of the lucky ones; I had survived. The landing on Saipan was a very traumatic experience for anyone who hit the beach. My mind and body were stretched to the breaking point. As we neared the beach in the grey of the morning, some words started coming to mind. I wrote them down on an old algebra book that I always carried like a Bible. It doesn't matter if they are good or not. I will forever leave the island, and the many sad memories of World War II behind with the words I jotted down on that D-day so many years ago.

> Our time has come to hit the beach.
> The tight defenses, we will breach.
> Men will die, and what a shame.
> Some will live to reap the fame.
> The gates of hell are drawing nigh.
> Lord give me strength, to live or die.
> Tis a bright silvery morning,
> Reminds one of a little fawn,
>   but twill soon turn to mourning,
>   for the morn is battle's dawn.
>   Lock and load

Korea would be a new kind of combat experience for me. The worries and fears of World War II were behind me. It was now time to start a new series of imaginary worries and fears for what lay ahead in Korea. These thoughts cannot be avoided, so you have to expect them and do your best to cope with these mental parasites. Because of my many previous close calls, I was certain the odds on my living through any more battles or skirmishes would be far less for me than for anyone else on the ship. The closer we got to Korea, the more these mind-boggling thoughts continued to build. This would be a different kind of war, because the stress, worries, and fears would not be dispelled instantaneously, as they had when you hit the beach on the islands. This would be a continuous operation, so these annoyances and anxieties would continue to incubate and bug your sanity as long as one enemy soldier lurked in front of your position. Your previous experience in a combat zone helped to insulate, and isolate, your mind from the realities of what lay ahead to some extent, but part of your mind is always in reserve, ready, and braced to accept the inevitable.

I was older now, and I was hoping I wouldn't have to run any more gauntlets of fire. Doing the boondocker ballet is very exhilarating, but only if you survive.

As we neared Korea, I wondered, would this be the end for me? I had been around so much death and destruction that I wasn't expecting any special considerations or mercy either from the enemy or the Almighty God. Lady luck will only embrace a combat trooper for a limited number of close encounters. I was certain I had used mine up. I had seen lady luck desert many men when the chips were down. Would she now desert me when the odds were stacked against me? These were some of the many thoughts on my mind as we dropped anchor in Pusan Harbor in 1950.

Korea

After our landing in Pusan, we disembarked and were moved into a barbed-wire enclosure called a stockade. I don't know if the wire was used to keep us in, or for the security of the camp. It probably worked both ways. We only stayed for a couple of days, and it was impossible to get outside of the barbed-wire perimeter. This was almost like a prison camp, and I was glad when we boarded trucks and headed north. Everyone got a cold weather sea bag except me. I reported mine missing. I was told to report it to the CO in whatever unit I was assigned to. I knew this would be a useless complaint because most marine units didn't care to be burdened with extra supplies, especially since they had to move quite often.

We boarded trucks and headed north through the town of Pusan. The streets of the town were crowded, and it looked like it would be a good liberty town, but you didn't see any servicemen on the streets.

We made one very lengthy ride through the country, passing flat farmlands and rice paddies, always with the mountain ranges in the background. I was fascinated by the tall rock wall that stretched for miles. It is still being constructed, as it has been for ages. Rocks and stones removed from the farmlands are carried by hand and placed on the wall. This appeared to be a continuous chore when working the fields. It was a good and practical solution for disposing of the rocks. The wall fence was about three feet wide and six feet high. This must have been a community fence; it appeared to go on endlessly.

After the long ride north we finally arrived at the headquarters of the First Battalion, Eleventh Marines. I was assigned to B-1-11 and spent the night on a hard cot with the second gun section. The next morning all the new NCOs were interviewed by the battery commander. The battery commander congratulated me on my past combat record and assigned me as a forward observer scout sergeant with F-2-5. This surprised me as one of the other sergeants had already volunteered for the job.

The battery supply didn't have any cold weather gear, but I did get a used sleeping bag, one flannel shirt, one pair of long underwear, two pair of wool socks, and an automatic M-2 carbine with two-thirty round magazines. I was now ready to join the marine infantry on the front lines. My scout sergeant's job was to direct artillery support for the infantry company at the discretion of the company commander.

The following evening a jeep picked me up and took me to the company CP of F-2-5. The rest of the FO team was already with the company. My communication sergeant was Sgt. Schull, a Kansas man. His wiremen were Pfc. James Didier and Pvt. Gentile. My radio man was Cpl. Carrol. The company had been in this location for a couple of weeks awaiting orders and making daily patrols. The next evening the company had a staff briefing. Lt. King, my FO officer, made the briefing and advised me to make sure the FO team and all the equipment were ready to move. Our orders were to move out at first light as the assault company spearheading the push north.

I was very good at reading a military grid map; this was probably one of the reasons I was selected as scout sergeant. Map reading is a must. I received a map of the area, a compass, and field binoculars with a mil scale built in. These accessories with the radio man were all I needed to call and direct fire on the enemy. Many times these commands are transmitted and relayed back to a battalion liaison unit. They, in turn, relayed the fire mission back to the battalion fire direction center. I didn't know at the time that Cpl. Carrol would prove to be the best radio man I would ever have while in Korea. He cared for that radio like a mother hen takes care of her brood. I believe that if a hand grenade had landed among us, Cpl. Carrol would have thrown himself on top of the radio. Without this radio we were out of business.

Lt. King and I fired many and various types of fire missions while we were together. Some were as preparation for an assault on a hill or mountain by the infantry units. He taught me many important things about calling and adjusting artillery fire. The most important thing was to always fire night defensive fire. You did this every time you stopped, even before you had dug a foxhole. It didn't matter how tired or worn out you were after a long and tedious march. We called this night defensive fire, and it was insurance against any possible enemy attack. We would be able to call the preregistered artillery concentrations in front of our positions during the night. To accomplish this we would scout around and locate the most vulnerable points of the perimeter, then adjust artillery fire to hit these approaches to our positions. For each mission we were given a concentration number. If we were attacked during the night, all we had to do was request that particular concentration number and get immediate fire on the target. If you wanted to survive a nighttime assault, this artillery fire could be a lifesaver for all. When Lt. King left, I continued on this practice and taught it to every officer who was sent up to get some experience as a forward observer. Sometimes these officers were only with me for a couple of weeks, and at other times for a month or two.

I just assumed that all officers were taught map reading in officer candidate's school, but most of the ones with whom I came in contact had a very limited knowledge of map reading. I always kept track of my position while on the march. When we stopped I relayed our position back to the battalion. This was necessary to eliminate the possibility of our own artillery units firing intermittent and harassing fire on our positions during the night.

When the company was on the move, we were up and ready to move out at the crack of dawn. We would be on the march, up and down hills and mountains until you felt like dropping. Sometimes it meant stopping after dark, only to receive word that there was to be no smoking and no building a fire. How many hills and mountains did we climb? I have no idea, but I do remember some of them and some of the action that I was personally involved in.

On arriving on the top of one hill, the company was subjected to small arms fire from an adjacent hill. The company returned the fire, and Lt. King fired a couple of rounds of artillery on the hill. One of our platoons prepared to assault the enemy position when a Korean soldier appeared with his hands in the air. After the order cease-fire, a squad of men were dispatched to escort the man back to our position. He told the CO that when the artillery fired, all of the soldiers had run into one bunker. The skipper called for the rocket team. While the rocket team set up and bore-sighted in on the bunker, the Korean waved and indicated they were sighted in on the wrong bunker. He got behind the rocket launcher and did the bore-sighting himself. When he was satisfied, the rocket crew loaded up and fired two rounds, hitting the bunker both times. A squad of men was sent out to check the bunker. The Korean was right; they found about a dozen dead soldiers inside the bunker. This was a thoughtless way for a combat buddy to treat his comrades. The hill secured, we continued our sweep to the north.

We stopped on a mountain for one night, and the company sent out a reconnaissance patrol the next morning. Lt. King and I went with the patrol in case artillery support was needed. About three miles in front of the lines, we encountered an oriental patrol on a trail down below us. It didn't appear to be the ROK army, but we couldn't be sure. They waved at us, and we waved back. They didn't fire at us, and we didn't fire at them.

They were sitting ducks and they knew it. We had caught them off guard. They had to act as friendly troops. If they had acted suspiciously or had started to run we would have opened fire.

Every evening when we stopped, the Korean laborers brought our rations up for the following day. The rations were strapped to their backs. As soon as they unloaded, they went back down the mountain. We got our own water by going down in the valleys and finding a small stream. Sometimes we had to break ice from the top, and other times we might have to dig a small hole and let the dirt settle before filling our canteens. When we were on the move, each man would empty his old water and refill it from the valley streams. If you were smart you tried to conserve enough water to make a cup of coffee once you were back on top of the mountain.

One time when we came off of one mountain, we filled our canteens with the only water we could find. It was dirty and foul tasting. Later in the day while on this march, we started up a valley beside a cool running stream. We all poured the old water out, took a big drink, and refilled our canteens. We continued on for about 100 yards when we came to a couple of swollen enemy bodies lying in the water. We went upstream above the bodies, emptied our canteens, washed our mouths out, and refilled them with fresh water. Another 100 yards upstream we found some more decaying bodies in the water and had to repeat the whole process all over again.

You carried your toilet with you. It consisted of an entrenching tool and a flat pad of toilet paper. This was generally protected in a cellophane bag along with your powdered coffee, cream, and sugar.

The C rations weren't as bad as the ones we had had in the Pacific in World War II. They were very good in comparison, that is, if you had time to build a fire and heat them properly. Most of the time I didn't fire night defensive fires until well after dark. I would then dig a foxhole in the dark before eating. Most of the time it was well below freezing, but no fires were allowed and you could only smoke undercover in your foxhole. More than once I have opened a cold can of C ration in the dark and, without knowing what it was, taken a big bite and started chewing, only to discover I was chewing on a blob of cold grease that had settled to the top of the can. If you were a coffee drinker like me, you did without until daybreak and hoped we would be allowed to build a fire. Many times I have spent most of the day going down the mountain, filling my canteen, and then climbing back up the mountain. I would then fill my canteen cup full of water and set it on or near a fire to heat. Once it was hot enough I would then dissolve a bag of powdered coffee in the liquid. At times when it was about ready, it would tip or fall over, spilling the contents. To repeat this for a cup of coffee was to use the rest of the water in your canteen. When I had a spill I did without rather than be on the mountain without any water.

We had many small skirmishes and confrontations during the first couple of months in Korea. We would secure one objective and move on to the next. Some objectives would be given to us, and others would have to be secured with a fight. Always stopping after dark and moving out at the crack of dawn was the norm.

One night I put my sleeping bag on the wet ground and was awakened early with orders to move out. the wet dirt was frozen to my sleeping bag. My hands were numb, but I did my best to roll up the bag along with all the frozen dirt. It was big and weighed a ton. My back was about to break as we hiked up and down the mountains. I was glad to see the sun come out and warm things up. Every chance I had, I unrolled my sleeping bag, kicked the icy dirt and mud off, rerolled it, got it back on my back, and then double-timed to catch up with the company. It only took this one time for me to learn that when you unfolded your sleeping bag for the night, place it down on pine needles, limbs, anything, but never on the bare ground.

One incident of wrong coordinates being called back to the rear resulted in a number of casualties for Fox Company. Most of the time a lieutenant checked with me, and then consulted the skipper before calling the coordinates of our position back to the supporting units in the rear. It so happened this one evening the coordinates were called back without consulting me. I had the right coordinates because I was upon the ridge line firing night defensive fire until well after dark. We had been on the march all day, and I was worn out after firing night defensive fires. I ate a can of C ration and crawled into my sleeping bag on top of the ground to try and get some rest. This was the first night since I had been in Korea that I hadn't dug a foxhole.

About ten o'clock our hill was hit by a barrage of artillery fire. This was friendly fire coming from our rear. The first platoon was bringing its killed and wounded back through our command post while I had the radio set up to get a check fire from all artillery units. I also relayed my co-ordinates back to battalion. I got a check fire from all marine artillery battalions. About the time the first platoon had its casualties removed to the rear, we received another barrage. Without a foxhole, I smelled gun powder, dodged tree limbs, and hugged the deck. I again called back to battalion with this information and was told it was an army unit doing the firing. Capt. King, who had taken over the battalion liaison after he had left me, advised me he had located the army unit and had now completed a cease-fire from the army. Fox Company was removing more wounded from this new barrage. I was lucky not to have been one of the casualties. Capt. King said that the reason for the artillery fire was that the company commander had called back the wrong coordinates for his company. I was told I was very lucky to have called back the correct coordinates when I did, because the Eleventh Marine, Artillery Regiment had been just one hour away from firing a regimental Time on Target barrage on our hill. This was when all guns fired simultaneously, using time fuses, on a suspected supply line or hill without warning, trying to catch the enemy off guard. These rounds would have exploded twenty feet above the ground, throwing shrapnel downward. This would have devastated our hill and annihilated most of Fox Company, including myself. I always wondered why the ability to read and orient a map wasn't a mandatory requirement for any officer leading a group of men in combat. I hope by this time pocket computers are available that will aid a novice map reader. This would reduce the number of many lost patrols and lost souls wandering around in the boondocks.

Ambush on the Pyonggang

When you started climbing up a mountain in Korea, you may have been clean and dry, but in trying to maintain your balance and footing you were continually grabbing at limbs and shrubs to keep from slipping and falling on your butt in the mud, snow, slush, or ice. You soon became covered and saturated with mud. You then gave up trying to keep clean, and you trudged along like a zombie—falling, getting up, and repeating this over and over—as you climbed and descended the various mountains. Strangely, the roughest part was not climbing up the mountain but walking down the mountain. This could be very painful to your knees. As dirty and as cold as you were, you crawled into your sleeping bag at night without washing or removing your shoes. I never zipped up my sleeping bag. You never knew when you would have to come out of it very fast. If you got any sleep, it would only be for one hour at a time. If your legs were straightened out they started screaming for relief. You would have to bend them up, doze off for another hour or so, until they started hurting so bad you had to straighten them out again. This routine was something you lived with every night. You never slept all night without some kind of interruption. My back injury from the blast on Saipan always hurt, but I was always able to get some relief from the small codeine tablets from our corpsman. They alleviated some of the pain, but I had to use them sparingly, as they made me drowsy. Most of the time we had fifty percent watch at night. If an attack seemed imminent, we would have one hundred percent watch.

One mountain was so big it took us a week to reach the top. This was not a straight-up climb, but you went up and down smaller hills as you continued the ascent to the peak. We climbed numerous mountains; if there was no opposition we continued down to the other side and on to our next objective. Since we had spent the better part of one week climbing this mountain, we got so far ahead of our supply lines that the laborers were unable to reach us in one day to deliver our rations. We finally descended this mountain into a valley. We could not advance any farther because of a river blocking our path. It was late and we dug in for the night. We had no sooner finished digging our foxholes when we started receiving mortar fire from across the river. You could hear the sound of the mortars when they left the tubes. With this information I was able to calculate the distance, and using the map I was able to pinpoint their position. I called for a barrage from a 155-mm battalion, or Long Toms. This was the only artillery that could reach the enemy coordinates. After the barrage, the mortars were silenced, but not before our corpsman was badly wounded.

The CO called for a helicopter to evacuate the wounded corpsman. The helicopter loaded the corpsman aboard. On one side of the helicopter in another stretcher were ten cases of C rations. Our CO had the pilot unload the C rations under protest, as they were destined for another unit. Everyone in the company got a can of C ration. It didn't matter what kind you got, you were so starved anything would have tasted good. This was the first food we had eaten in three days.

We received a briefing that once we crossed the river, we would be relieved by the army. This was good news, as we had been on the move every day for over two months without relief. The river was too deep to ford. Amphibian DUKs, or ducks, as we called them, would come up the river and ferry us across the following morning. Every time the ducks got in our vicinity, they received enemy fire. They decided to wait for us around a bend in the river. We had to climb a very steep hill and go down the other side to the river and the waiting ducks. It took us most of the day to climb the steep hill, and it was getting late when we descended the other side as fast as we could. We had to zigzag on our way down because it was so steep. Once in the gravel bed of the river, we waited for the two ducks to come over and ferry us across. Both ducks were on the far side when all hell broke loose. Heavy small arms fire kicked up sand, gravel, and rocks all around us. Everyone scattered to find cover. I made a headlong dive behind a log on the bank of the river. Bullets crackled all around me, still kicking up gravel and rocks. I thought about calling in artillery fire, and I looked over the log to see if I could determine where the enemy fire was coming from. When I peeked over the log, I was looking into the face of Lt. King. He said I was on the wrong side of the log; the enemy fire was coming from behind me. I made a quick flip over the log. The enemy had waited in ambush, until they had us outnumbered, before striking. I fired my carbine at the hill we had just descended, but we didn't have the firepower to stop the attack. Just about the time I thought we were fighting a losing battle, I heard a *blub, blub, blub*. Looking over my shoulder, I could see one of the ducks in midstream firing one of their mounted .50-caliber machine guns. This fire really churned up a storm on the side of the hill. It continued to fire as they ferried us across. I had given the enemy a perfect target. I don't know how they missed.

We were now across the river and officially relieved by the army. There was no transportation to take us to the rear, so we started walking. The Second Cavalry continued to move up in trucks. They jeered at us and called us bellhops. We yelled back at them and asked if those were saddle blankets on their backs. We walked all that evening and all night, arriving at the position of B-1-11 just at daybreak. I was worn out, but the mess tent was getting ready to serve the morning chow. It was below freezing, but I was starving. I broke out my mess gear and went through the chow line. I ate twelve boxes of dry cereal, using powdered milk that had been mixed in cold water. Even with all the lumps in the milk, it was still the best meal I had eaten in almost three months.

The battery supply still didn't have any cold weather clothing. The supply sergeant was the one who had volunteered for the job they gave me. I had been on the march for three months with only my dungarees and field jacket. All the other men had heavy alpaca coats, caps, vests, and heavy shoepacs. One thing I really missed was not having any gloves. I was still wearing my old rawhide boondockers. We had waded across streams in freezing weather, and I think my boondockers were much better than the shoes or snowpacs that the other men wore. I was only back with the battery for two days when my FO team was ordered back to the front lines. The supply sergeant was nice enough to give me his alpaca coat before I left. The sleeves were a little short, but this was a lot warmer than my old field jacket.

Two days in the rear were not much of a rest, but it was a welcome break from the front lines and C rations. The battery trucks took us back to join F-2-5 at the front, to continue our push north to our new objective.

A Test of Endurance

The Korean mountains and hills all look alike, but after you have climbed many of them you begin to note the peculiarities and personalities, the shape, roughness, smoothness, and the sheer steepness of others. Anyone not versed in map reading could easily become confused with the similarities. If you didn't keep a running check on your map as you advanced, you would be lost when you stopped. This would cause you to overshoot your objective or fall short of achieving it.

At times when we were briefed as to our mission, generally a mountain or hill would be marked off on our map as being the objective. I noted that on my map, and kept a running check as we climbed up and down the mountains and hills. I always marked or noted any prominent landmarks during our march, so when I stopped I would know the exact coordinates on the map. This was vital if I was to fire artillery on the enemy. Many times officers became confused or even lost because they didn't keep a running check of their positions. For some reason they were too bashful to ask what our present coordinates were.

On one march we had been on a constant move since daybreak. As usual I kept my map up-to-date. Just at sundown we stopped on a mesa, and the CO said we would camp on it for the night. We were short of our objective, but a tempest was bearing down on us with dark, rolling clouds. We raced to get our pup tents, or shelter halves, erected before the onslaught of the storm. The storm hit us with a vengeance and harshness that made me thank the good Lord for letting us get our shelters set up in time. Strong, sweeping, icy winds, and a combination of rain and sleet was released in a torrential and frigid deluge. It would be impossible to fire night defensive fires under these conditions.

Ten minutes after the storm hit, I heard the word yelled to stand by to move out. I got the word that we were about five miles from our objective. I already knew this, but I thought we had stopped because of the storm. We packed our equipment and moved out with the company. Everything was waterlogged and, therefore, weighed much more than it normally would have. Moving out in this kind of weather with all this weight was sheer torture. Slipping and sliding up and down the mountains along the rugged mountain path reduced your clothing to a cold, freezing, oozing mud. It was grueling to climb up, grabbing at the limbs or the shrubs or whatever you could get hold of while groping in the darkness, and then slipping and sliding down the other side of the mountains. With the driving rain and the darkness you felt enveloped or isolated in a quagmire of muddy mountains, waiting to dissolve in one big avalanche of ooze once you lost your footing or slipped. I was becoming a walking robot; the numbness of the cold had seeped into the core of my body. I had to keep moving or my feet would freeze. I could feel ice crystals forming in the toes of my boondockers. I have no idea how many men fell down on the side of the trail, bawling like babies. These men had to be led back down the mountain to an aid station. The distance to the aid station was probably a lot farther than the distance to our objective. I wasn't about to give up. I had come a long way, and I didn't intend to be beaten by the elements. Hang on for each mile. Every hour extends the elements in your favor. You must hold onto your sanity until the sun comes up. Once it breaks over the peaks, your senses and mental balance are once again in tune with time and the universe. You, and the men beside you, have just overcome another obstacle in the quest to secure your objective.

We arrived on the peak of our objective some time in the early morning. The icy rain was unrelenting. It was so dark, and the rain so heavy you couldn't see your hand in front of your eyes. No lights or fires were allowed, so we groped around in the muck trying to find a foxhole or a place to dig one. While groping around, my FO officer and I stumbled onto an old foxhole. We set about stretching a poncho over the hole for protection against the pouring rain. We crawled in, and sat down in waist-deep water. This slight respite from the cold was all it took for sheer exhaustion to take over. My mind and body ceded the last trace of resistance to the elements. The men in F-2-5 had just survived a grueling test of endurance beyond belief.

We woke up to a bleak, grey, and foggy morning as we climbed out of our muddy foxhole. The water had seeped out, and we could tell there were boards in the bottom of the foxhole. Someone had a fire going, and we spent over an hour getting warm and trying to dry our clothes out. Our clothes gave off a foul stench even though we had undergone a change in mud and dirt.

We were on the peak of a mountain. The fog had settled to just below our peak, and we looked across a vast ocean of white fog to other peaks miles away. It was like being stranded on a small island, looking across the soft ocean waters to the other islands miles in the distance. It gave you a weird feeling, like sitting on top of a small island after a big flood.

Once we had warmed up and had gotten the chill off, we decided to investigate the boards in the bottom of our foxhole. We thought it might be hiding an entrance to a bunker. We took our entrenching tools and pried up the boards in the base of the hole. The stench was overpowering. The boards had been placed over a dead body. We had just spent over eight hours sleeping in water on top of a decaying body.

We had reached and secured our objective. We had survived the ordeal and met no opposition. We packed up and moved out to our next objective.

A Break from the Front Lines

We were always climbing hills and mountains; if they were occupied the enemy engaged us in a fire fight for a short period and then withdraw, only to catch us coming up the next hill. Each hill or mountain was a gamble. After one of these fights, I procured a nice sleeping bag from one of the evacuated wounded. It was good to discard my old one. Even with the newer bag, I still climbed into it every night with my dirty boondockers. Never zipping up, no matter how cold it got. I felt helpless and always had a fear of being bayoneted while trussed up in a sleeping bag.

One evening after the marine unit on our left flank had engaged the enemy, they started evacuating their killed and wounded back through our position. Most of the bodies were burned and charred beyond recognition. Almost the entire company had been wiped out with napalm. This was caused by our planes using napalm in a misdirected air strike. The marine mortars had fired smoke rounds to mark the enemy's position, but before the circling planes had come in for the strike, the smoke had drifted back over the marine company. The planes made direct hits on the marine company and since the marines had just arrived in this position, none of them had time to dig in. Most of the company was wiped out because of this error. I always felt a sense of rage every time something like that happened, but they continued to happen, and there was nothing anyone could do about it.

Over a period of time we had lost a lot of men in our company. We were pulled off the front lines and went into reserve to await replacements. Our stay was short. For some reason we moved out the next morning on our way back to the front lines. We started out, winding our way up a long valley. At one point we turned and started our climb up a mountain. At this time, I didn't have a clear-cut knowledge of where the lines were located. What were called the front lines were never in a straight line. I did know the coordinates of my position on the map. As my FO team and I were about halfway up the mountain, mortars started exploding down below us. I hoped the end of our company had cleared that point.

This was not our luck. Someone yelled for the corpsman, and I knew we had casualties. They called for me too. I dreaded making the long trip back down the mountain, only to have to climb all the way back up later. It was a very steep incline that made it so hard to climb. When I reached the point of impact down below, dead and wounded were scattered all over the trail at the base of the mountain. Platoon Sergeant Jeffcoat and one of his squads had received a direct hit. Most of them were dead and were being evacuated. I was asked to make a crater analysis and to get some artillery fire on the enemy mortars. While the killed and wounded were removed, I dug into the smoking craters that had just been made by the mortars. I dug down to the fuse in each crater, and without disturbing the fuses, took a back azimuth from each one. I made about six, and by using these, I was able to pinpoint a most likely position for the enemy's mortars. I fired the 155-mm Long Toms on these enemy coordinates, and we never received any more hostile fire. Before leaving the site, I filled in one hole and buried someone's leg that had been severed and overlooked during the evacuation of the casualties. Any time you lost good men like this, it only increased your determination to locate and destroy the enemy with any means possible, and without compassion.

Another time we had lost men and were down to a very weak company. We were pulled off the front lines for a short break. By now the hope of receiving replacements for the men we had lost almost deserted us. We left the mountains about noon and marched for most of the evening. It was late when we set up behind a small hill in some rice paddies. The rice had been harvested, so only the stubble was left. Being behind this hill should have shielded us from the enemy, or so I thought.

I went to an old Korean house beside the rice paddy and found a wide six-foot board. When I laid this board on top of the cut-off rice stalks, it served as a makeshift innerspring mattress. I spread my sleeping bag out and prepared to get a good night's sleep. I didn't have to fire night defensive fires, so I was going to turn in early. I really thought we were safe, so I neglected to dig a foxhole. For the first time since I had been on the front lines, I took off my old dirty boondockers before crawling into my sleeping bag. Also for the first time ever I zipped it up and laid on the springy board. I felt relaxed and fell asleep immediately.

About one o'clock in the morning our Garden of Eden exploded. Artillery and mortars exploded all around us. Being zipped up was like a nightmare; I rolled over and over, fighting my way out of the sleeping bag.

I ran about seventy-five yards through exploding rounds to the next tier or bank of the rice paddies that offered the only means of protection. It was dark, and the only way I could see the bank was from the light flashes of the exploding mortars and artillery shells. As I made the headlong dive for the bank, something struck me on the side of my head.

The force of the blow knocked me sideways. The whole side of my head hurt. I thought I had been struck by shrapnel, and it felt like mushy brains were oozing out from the side of my head. I just knew it was a fatal wound. In the light of the exploding rounds, I wiped the sticky goo from my ear and the side of my face. The whole left side of my head was a gooey mess. It appeared to be blood, and I began to feel very weak and on the verge of going into shock. I felt my head to see just how badly I was injured. I wiped all the goo off and found it to be nothing more than mud. A large clod of mud from the exploding rounds had slapped me beside the head. My ear was still ringing and full of mud. I was still a casualty. I had made the wild sprint, doing the boondocker ballet across the rice paddy, without my boondockers. I had run over the sharp rice stubble with my bare feet. The bottoms of my feet bled and looked like raw hamburger. The corpsman gave me some salve, but it still took over two weeks before they were back to normal. I continued to walk on them, all day, up and down the mountains and across icy streams.

It seemed no matter how experienced you became there were still lessons to be learned the hard way. After this experience, I never again completely zipped up my sleeping bag or removed my boondockers before crawling into my sleeping bag. On the front, or in reserve, I was never again guilty of not digging a foxhole.

We went back on the front lines after being off for only two days. Not much rest, but we had been extremely lucky. Over seventy rounds of artillery and mortars exploded in our bivouac area and didn't cause any serious casualties. The enemy had missed. They had given us a new lease on life, and a chance to fight yet another day.

Slaughterhouse on the Soyanggang

We arrived on top of a huge mountain late one evening overlooking Slaughterhouse. Lt. King and I fired our night defensive fires on the most likely approaches to Fox Company's position. Slaughterhouse is obviously not a Korean word, but that was the name on my military map, so I assumed it was the name given for this area for some reason unknown to me.

We were on this big mountain looking down on smaller mountains. Looking north was a wide valley with mountains on both sides. A stream meandered down through this valley and emptied into the river at the base of our mountain. There was a large river, the Soyanggang, running from left to right if you were looking up the valley. Everything across this river was controlled by North Korea. About halfway up the valley were a few old houses, and between the houses and our mountain was a road, also leading from left to right. A small, one-lane, rickety, wooden bridge spanned the small stream. We were up so high we could look down on everything, like the zigzag trenches that ran from the top of the mountain on our right of the valley, all the way down to the road. With my binoculars, I could see the weapons in the trenches. You could make out the machine guns and, at the base of the mountain where it joined the road, was one machine gun and one rocket launcher. All were in place with ammunition, ready to fire. We were unaware that we had a ringside seat for a number of events that would unfold over the next few days. It was as if we were sitting in the top balcony, watching a large stage presentation. The first act began the following morning.

I had tried to register an artillery concentration on the trenches but was told that these coordinates were in the Sixth Marines area, and I would not be allowed to fire in their zone.

At eight o'clock the next morning the Sixth Marines were moving out, either to patrol or to try to take the mountain with the trenches. Two tanks led the parade; the troops moved behind the tanks. They were moving from our left to our right. The North Koreans made a wild dash from the bunker on top of the hill down and into their trench positions. The enemy covered the bridge and the road. If they had given me permission to fire, I could have blasted them out of the trenches, but I was still denied permission to fire into the Sixth Marine zone. This was the rule and they stuck to it, even though I advised our battalion CP that I could well observe the enemy and the Sixth Marines.

We could only sit and watch as the Sixth Marines moved up closer to the bridge. The machine gun and the rocket launchers that covered the bridge and the road were manned and ready to fire. The marines approached the bridge, the only crossing of the deep stream bed. All enemy guns and mortars were registered in on this bridge. We sat helplessly and watched, wishing we could warn the marines. The troops and the tanks came under heavy fire as they neared the bridge. One tank was knocked out of action. We finally contacted the tank commander and directed him to the machine gun positions up in the trenches. The 90-mm guns from the tanks left gaping holes where these guns had been located. We were not finished directing the tank when the Sixth Marines withdrew, leaving the damaged tank. We requested and received permission to direct intermittent artillery fire above the tank during the night to keep the enemy away.

The next morning the Sixth Marines called for an air strike. While this air strike was in progress, the Sixth Marines retrieved their tank. We worried about the air strikes. We were so high, that if the planes didn't stop firing in time, as they pulled up some of the bullets would land in our area on our mountain. The North Koreans stayed in a bunker on top of their mountain. When the air strike was over they made a wild rush down through the trenches to man their positions. We had to go all the way back to the division switchboard with our communications to make contact with the commander of the air strike. We pointed out the red clay-topped bunker on top of the hill as being the main bunker. At the same time we asked the air strike commander to reverse their approach to the trenches, so that we wouldn't receive any stray bullets when the planes pulled up from their dive.

This second air strike was perfect. They hit the right bunker with napalm, rockets, and bombs. One plane flew a zigzag pattern along the trench line from the base of the trenches all the way to the top, churning up dirt all the way. I wouldn't have believed this possible if I hadn't observed it with my own eyes. Later the planes made one more strike, doing even more damage, but the enemy still had enough men to man their trenches. The rest of that day remained quiet.

How lucky can a person be? Two army soldiers may never know or realize just how lucky they were that day. The action started just at daybreak the following morning. Someone yelled that the Sixth Marines were moving out. I grabbed my field glasses or binoculars and ran to the top of the ridge line. It wasn't the marines, but a lone army truck with a big white star on the side of the door. About a dozen laborers rode in the back. They had taken the wrong road and were lost.

The driver came to the rickety bridge, stuck his head out, and then slowly eased his truck across the bridge. I checked the trenches. They were empty, but you could still see weapons in place. We had no way to either warn the soldiers or turn the truck around. The truck came to the base of the trenches by the road. The driver could have gotten out of his cab, walked about ten feet, and picked up the enemy's machine gun and rocket launcher. The driver finally decided he was on the wrong road. I kept saying, "Get out of there fast; move it." I held my breath as the driver pulled forward, then backed up, then pulled forward again, trying to get turned around. He took his time, but he finally succeeded in getting the vehicle turned around and back on the road going back the way he had come. As he neared the bridge, I could see the North Koreans sprinting down the trench line from the bunker on top of the hill just as fast as they could run. The driver stopped and then eased his truck over the bridge and back down the road. They were now out of range of the enemy's weapons. I was breathless and really tensed up. These two army guys were very lucky; they had caught the enemy unprepared. What a relief! Not only for me, but for the soldiers.

The Sixth Marines called for another air strike. This time it was different planes; again we had to go back through the division switchboard to talk to the air strike commander. Once again they were right on target and must have caused the enemy many casualties.

I do believe we had set some kind of a record for forward observers. We had directed an adjusted just-about-everything. This included a precision fire mission on one of the houses way up the valley. Enemy soldiers were observed going into one of the houses just at dusk. We still had time to fire the mission. In the fire for effect the house was blown down. Only a couple of soldiers were seen running from the ruins.

Lt. King received his promotion to the rank of captain and left Fox Company for the rear. I was now soloing. Any fire missions from this point on would be directed by myself until another officer was sent up to take Lt. King's place.

The following morning the trenches were deserted, and we received orders to move out. Two hours later we crossed the Soyanggang River. We forded the river and came out on the road at the base of the trench line, where the machine gun and rocket had been located. We moved out along this road without meeting any resistance. After three miles we turned and went up a valley leading north. Late that evening we arrived on top of a hill overlooking a small river. Across this river was a single-lane road winding down from our left and crossing the river to our right. I wanted to fire my night defensive fires on the road and on the hill just in front of us, beyond the road. This hill was about the same height as the one we were on. I was instructed that all friendly troops were withdrawing from across the river, and I wouldn't be able to fire any missions until all friendly troops were back on our side of the river. I could only sit and wait. Trucks and many foot troops were coming back across the shallow part of the river. I waited until well after dark before I fired my night defensive fires. I registered one concentration across the roadway to my left, and another one on top of the hill across from ours. Dog Company was set up on our left flank, and after I finished my night defensive fires, it suddenly occurred to me that I didn't hear Dog Company's FO fire their night defensive fires. Maybe they thought it was too late.

The troops' withdrawal was an indication that pressure was being exerted from the enemy. The CO called for 100 percent watch during the night. No one would get any sleep. Over a corps of Chinese troops were massed in front of our lines, and we expected a huge all-out assault during the night. I sat up all night, leaning with my back against my foxhole holding the field phone, ready to call in artillery fire once we were attacked.

It was about one o'clock in the morning when the Chinese started banging on pots and pans and blowing their bugles. It sounded like a New Year's Eve celebration. Within an hour Dog Company came under attack; the FOs did not fire any artillery. My concentrations weren't close enough to give them any help. We were already braced to repel any assault against our position. Dog Company continued a heavy fire fight for the rest of the night. When daylight came, Dog Company started evacuating their killed and wounded back through our lines. I don't know how many enemy were killed during the night, but no matter what the odds, Dog Company had held its position.

The next morning I had Carrol, my radio man, set up his radio and get a check with battalion. This check had just been completed when I looked across the valley to the hill in front of us. It was covered with enemy troops. The hill was black with soldiers who were shoulder to shoulder. I stood on the ridge line and looked through my binoculars at all the troops massed on the hill. I observed two high-ranking Chinese officers with a white officer in a different uniform. All were standing together looking straight at me with their binoculars. I yelled to Carrol, "Fire mission!" He relayed this back to battalion. I requested concentration Charlie 24, with battalion two rounds of variable time in the fire for effect. Thirty-six rounds of 105-mm ammunition with time fuses that exploded twenty feet above ground, throwing the shrapnel downward, hit the enemy without warning. The barrage was right on target with my previous nighttime registration. The rounds exploded all over the forward slope and ridge line. Dead soldiers were knee deep, lying all over the hill, including the three high-ranking officers. Those not killed or wounded were running to the reverse slope of the hill. I added 100 yards and requested a repeat fire for effect. This time a few rounds landed on the ridge line, but the others landed on the reverse slope. It was a perfect mission. I observed over 800 bodies lying all over the hill, and possibly that many on the reverse slope.

Coming down the road in the valley from our left was a mass of troops that's hard to describe. It was like a giant close-order drill parade. Enemy troops marched shoulder to shoulder, from the river to the hill a mile away, stretching back up the winding valley as far as you could see. I requested my other concentration, the one I had fired across the valley on the road. I was told the artillery battalion had been secured and they were pulling back. There was no one capable of giving us artillery support. I told Carrol to secure the radio and to get ready for one hell of a fight. The enemy had more troops than we had ammo. This was a very helpless feeling, just sitting and watching a corps of Chinese moving up on you in a no-win situation. All anyone could do was take out as many of the enemy as possible. Either way you could be killed before it was all over. Our time had come at last. At the last possible minute the word was passed to move out; we had orders to withdraw. Just before I left the hill, I took one more look across the valley to the three high-ranking officers. Their bodies were still where they fell. I didn't see any movement on this hill. All the bodies I observed were lifeless. For these soldiers the war was over.

We came off the hill in single file. I was midway in the company formation. As I came off the base of the hill I passed four tanks that had started firing .50-caliber machine guns at the hill we had just vacated. I looked up at the top, and the hill was black with Chinese. I could see the last man in our company just coming off the base of the hill. We went over a mile back down the valley to the road leading to Slaughterhouse. We continued on down this road, with the sound of the tanks still firing as we passed the base of the trench line and the road. I've often wondered if the tanks ever got out of that valley. They had covered our withdrawal, but there was nothing or no one to cover them. They were still firing as we crossed over the rickety bridge on our way south. We marched for miles, but the tanks never caught up with us. We continued our march for two days and nights, stopping on the Quantico line. This would be our final protective line. We would dig in and hold this position at all costs. There would be no more withdrawing. Everyone was dead tired and worn out, but we dug in and set up our defensive guard before anyone got a break. We had moved so fast that we had taken the wind out of the Chinese offensive. The enemy had overextended its supply lines and was slowed to a crawl. We would have time to reorganize and start our push back to the north. It had been hard to pull back past many hills and mountains where we had lost good men fighting to take them. I was told not to worry about this, because we were not fighting for the real estate.

The reason for our sudden withdrawal was because the Chinese had broken through the ROK army and were trying to trap or encircle the marines. Whatever the reason, I was more than glad to leave when we did.

The F.O. Team Wiremen

It was routine just to march all day, starting before daylight and stopping at sundown. For some of the men a stop meant a break or at least a rest from eternally marching and hiking up and down the mountains.

For me, it meant making a fast survey of the area to locate the best places to fire my night defensive fires.

For my wiremen, it meant taking small drums of wire and laying a telephone line from our position on the front to the battalion CP in the rear. This was dangerous as my three wiremen were always exposed to sniper fire, and always in danger of stepping on some derelict land mine planted in or along the valley trails.

We always looked for a better method of doing things, and I always listened to any suggestion that would improve our duties.

One of the wiremen said that since he was good at reading a map, they could stay back with the battalion CP and lay the wire up to the front lines when I called back my map coordinates. I thought this was a good idea. This way they wouldn't have to hike all day, carrying the heavy and cumbersome phone and reels of wire. We decided to try this and see how it worked out.

Signs of the enemy were all around us. We had marched all day, expecting to be attacked at any moment.

That evening we stopped on top of one mountain. Directly across from us was another mountain a little larger than ours. We expected the enemy to be occupying this mountain. I called my map position back to the battalion, so the wiremen could start laying the land lines up to our position. I then fired my night defensive fire on the most easy approach to our defense perimeter. I had alerted the infantry that I had wiremen on the way up the mountain. It was well after dark when I completed digging myself a good foxhole.

I had the radio man set up the radio and call back to battalion to see what had happened to my wiremen. Capt. King, who was in charge of the battalion CP, said the wiremen had already laid the phone line up to our position and had returned. They said the field phone was strapped about waist-high to a tree on top of the mountain. I looked all over the mountain in the dark trying to find the phones, but I had no luck. I gave up and tried to get some rest.

We were up and started moving out before daylight. We went down a long ridge line and across a valley, and we started our ascent up the next mountain. We expected this mountain to be defended by the Chinese. One platoon scouted the peak and gave the word that it was all clear. When I arrived on the peak, the first thing I noticed was the field phone strapped to a tree. My wiremen had laid the phone line over two miles into enemy territory. If the enemy had been occupying this mountain it could have ended in disaster. Luck was with my wiremen on this mission. We scrapped that method as being too dangerous. From then on they stayed with the company.

These were very good wiremen, and I was lucky to have had them in the FO team for most of my tour of duty in Korea.

Retaking Lost Ground

One morning F-2-5 was suddenly pulled off the front lines. We boarded trucks and were rushed to another part of the lines. The enemy had broken through the Second Cavalry Division. We set up a defense across a valley, to be a buffer in stopping the enemy's advance.

We were reinforced by a platoon of tanks which were also deployed across the valley. These would be invaluable in case of an assault on flat terrain.

The first night we had a full watch, expecting to be hit by the Chinese at any time. All the valleys to our front were bombed for most of the night. Saturation bombing techniques were used, and we observed many dead bodies on our advance to regain the ground lost by the army.

The second day a battalion of Dutch soldiers came down the valley from the front. They had been trapped when the army had retreated. They had to fight their way out. Their commanding officer said this was the second time this had happened, and he was about ready to take his men out of Korea.

When the Dutch battalion received mail, many of them received boxes from home with Dutch chocolates. They were very generous sharing these candies with the marines.

Once the Dutch departed, we started our push North to regain the real estate lost by the army.

We climbed hills all one day and that night we stopped on a small mountain. I say small because we were looking up at an enormous mountain range in front of us.

One event happened that I have never been able to explain. I had fired my night defensive fires on the two most possible approaches to our position. One platoon was going to put two sentries out in front for a listening post. I explained that if I fired my night defensive fire the men might be endangered. We came up with an alternate idea. We would put the listening post back with the company and run phone lines out in front, and connect these to a sound-powered phone. This phone was tied about waist-high to a tree. Everything was quiet until about one o'clock in the morning, when the listening post alerted me. They could hear enemy voices coming over the sound-powered phone. I listened, and sure enough you could hear what sounded like a Chinese commander barking orders to his troops. I immediately fired my concentration on the ridge line. I could hear the *pop, pop* of the artillery rounds exploding as the sound came over the sound-powered phones. After firing we listened and all was quiet; you could not hear any enemy voices. We expected an attack at any time, so no one slept. After about thirty minutes voices were back on the phones; again I fired artillery and again the voices were silenced. Thirty minutes later the same thing was repeated. Early the next morning I went down the ridge line with the men to retrieve the phones. They were still tied to the tree and were not damaged. I was surprised not to find any dead bodies lying around, but then the enemy had a habit of pulling the dead off the well-worn path and leaving the bodies in the underbrush, rather than leave them exposed to be counted. I still believe the enemy was in the range of the sound-powered phone when we heard the voices. I can't come up with any other explanation. After this we were afraid to rely on this type of listening post and never used it again.

We moved up, and on arriving on the peak of the next hill, we found ourselves overlooking an army camp. All the tents were still up, and everything was left intact. The Chinese had blocked the only way out. The soldiers had departed on foot and left the camp in a hurry.

We continued on around this ridge line overlooking the camp and found a few dead soldiers lying in their foxholes. Only a few had stayed and fought to the death.

We continued moving up, retaking the lost ground. Moving along one ridge line, we observed a couple of old Korean houses in the small valley below us. You see them all the time, but one of these houses had a Korean man standing in front waving to us. A patrol was sent down to see why he was trying so hard to attract our attention. Dead American soldiers were lying all over the valley. About a dozen dead soldiers were lying in front of the house. They appeared to have been executed. Most of them were clutching the small pocket-size Bibles.

Inside the house were three badly wounded soldiers. They were wounded so badly that the Chinese had bandaged them and left, probably expecting them to die. One of the men was a Hawaiian soldier, and the Chinese were going to kill him, thinking that he was Korean, but they convinced the enemy that he was not Korean so the Chinese spared him.

Helicopters were called in to evacuate the men. When we brought the men out of the house, one soldier pointed to one of the dead soldiers and said that that was his executive officer. He also said the commanding officer was the first to desert the hill. Once he left, all the other men had scattered. All the Chinese had to do was run around the ridge line, trapping them down in the small valley. They didn't find the commanding officer among the dead and the last I heard they were still looking for him. Even on this ridge line we found some soldiers who had stayed in their foxholes and fought.

We continued our march and descended a large mountain into a wide valley. This valley had a long winding road that was supposed to be the escape route for the retreating army forces. The Chinese had cut the road off, trapping all of the vehicles on the road. To see this massive loss of so many and varied vehicles lined up and destroyed was a sickening sight. All were left burned and inoperable. Every type of conveyance imaginable was bumper-to-bumper and stretched for miles. We marched for most of the day and finally turned off from the valley. We still had not reached the end of the derelict convoy. The cost of replacing those vehicles would have been astronomical.

F-2-5 didn't have any vehicles; our transportation consisted of two legs per man.

All the bridges we saw in Korea had the centers blasted out either from bombs or demolitions.

Sometimes the engineers constructed a pontoon bridge, but most of the time our company waded across the rivers and streams, sometimes in sub-zero temperatures. At times when we waded across a stream or river we would then take advantage of the first stop to wring out our socks and replace them with the pair each of us kept under our pack straps. This was a must to keep from getting frostbite.

The two toes I had frozen in New Zealand still gave me trouble when my feet got cold. I still wore my old rawhide boondocker shoes, and I could make a change a little faster than the men with the winter shoepacs. These men seemed to be having as much trouble or more than I, so I wasn't in a big hurry to change my footwear.

I had two pair of wool socks. Every morning I would put on a dry or mostly dry pair. I would take the old socks and put them under my jacket or coat and place them under the straps of my backpack. The socks would serve as padding for the pack, and at the same time would dry out to some extent before being switched the next morning.

Casualties Mount

Seven months on the mountainous front lines is a long time, especially if you are living out in the open. I was pleased when our lines remained stable long enough that I could return to B-1-11 for a couple of days. I arrived at the battery in the late evening. The guns were set up ready to fire. Behind the guns were six-man tents with cots and a diesel stove. One supply tent was erected to serve as the mess tent. After eating I was assigned to one of the tents. I was sitting on one of the cots warming myself, thinking how nice it was to be out of the eternally frigid mountains. I had forgotten what it felt like to be inside any structure that offered protection from the elements. The CO had sent over a case of beer for me, and I sat up very late drinking beer and enjoying the luxury of the moment. I finally laid back on my sleeping bag and fell fast asleep. It was another luxury to wake up and not have the flap of your sleeping bag frozen solid from the moisture emitted from your breath. The rest was much too short, and I was off again to rejoin F-2-5. I felt a lot better and wore a clean pair of dungarees. I was unable to get a pair of long underwear, and I didn't dare discard my old ones. This garment was the only thing, I believe, that kept me from freezing to death. I still wore the same old boondockers that I was wearing when I landed in Korea. They were just about worn out, but they had to do, as I was unable to get any other kind of footwear.

A new F.O. officer had joined the team. This officer was very sharp and quick to learn. This was good; it would relieve me from some of the pressures of scouting and firing night defensive fires.

About the time I rejoined F-2-5, the company moved out with a mission to secure a couple of objectives. We started out in a wide valley that gradually narrowed and was split by a large hill in the middle. The company was going to go around this hill, but as an added precaution the CO sent one platoon up the hill to make sure it was not being occupied by the enemy. My radio man and I went with this platoon in case they needed artillery support. There were three ridge lines leading up from the valley floor which converged on top of the hill. It was like a pyramid with the three spines meeting. We climbed one, another one led up from our right, and the other led up on our left. About 100 yards from the top, the platoon got caught in a cross-fire from two enemy-occupied ridges and the top of the hill. Most of the platoon had been caught and cut down in the open and many assisting the wounded also became casualties. The man standing next to me turned white as a sheet. I grabbed him and eased him to the ground; he had been hit in the chest. He was dead before I laid him down. I picked up his M-1 and emptied the clip through the bushes in the direction from which the fire had come. Leaves were falling all around me, so I knew the enemy was getting close. The wounded had been hurriedly dragged and carried back down the slope of the hill.

With all the killed and wounded evacuated, and all of the men involved in taking them back, we were left with about a squad of men. The only protection we had was from a large rock outcropping on the razorback ridge. The enemy would toss a grenade over the rock. If it landed on one side of the ridge we would throw ourselves over to the opposite side until the grenade exploded. If the next grenade landed on this side we would throw ourselves over to the other side. This was repeated a number of times. It resembled a conga line, but I still called it the boondocker ballet. I reloaded and put a bayonet on the M-1 rifle. I was now prepared to charge the enemy.

Gunny Ski had been hit in the buttocks and was losing a lot of blood. Even though his pants were soggy with blood, he picked up a Browning automatic rifle dropped by one of the injured marines to banzai the hill. I decided to fire a battery of one round on top of the hill before we charged. My radio man was quite a distance behind me, and I had to yell my commands to him over all the noise from the firing. Either my radio man misunderstood me, or the battalion knew we were pinned down by enemy fire and had decided to give us added firepower. I'll never know what happened, but instead of the battery one round, they gave us a battalion one round. The center battery hit the top of the hill, but the rounds from the right battery started landing all around us. I danced to dodge the trees and limbs that were falling around me. The concussion from the exploding rounds actually caused my eye to jump out of its socket. I pushed it back in with the palm of my hand. My nose was bleeding, but after that barrage was over I was prepared for a banzai attack on the enemy. We charged out from the cover of the rock. I moved fast and passed by the two North Koreans who had been tossing the grenades over the

133

rocks. They were both dead in the bottom of their holes from one of the artillery rounds, with their burp guns laying on the edge of their foxhole. As the enemy departed the hill, they ran over the top, emptying their burp guns at us as we came up the hill. I emptied the M-1 clip at the running North Koreans as they crossed over the peak of the hill. When I gained the top, they were at the bottom, starting up the side of another hill.

I threw the M-1 down and grabbed the carbine from my shoulder. It was like a shooting gallery. I caught them in the open and emptied my carbine magazine at them in a crisscross pattern. I hit many of them before they scattered. Then I reversed my magazine and emptied the second magazine at the ones trying to pull the wounded into the underbrush beside the trail. The hill was ours. I didn't believe the Koreans had enough men left to launch a counterattack. Another platoon arrived, and the platoon leader ordered Gunny Ski off the hill and back to the aid station. We rejoined the company and set up our defenses for the night.

A few days after this encounter, another one of our platoons was caught in the open, and many were killed and wounded. The enemy was on a hill about 100 yards away and on the same level with our hill. The platoon was caught in the depression between these two hills. I wanted to fire artillery on the enemy but when I peeked over the hill from behind a tree I was immediately struck with dirt and splintered bark from the machine and burp guns. Once the wounded had been removed, the new F.O. officer and I fired a mission on the enemy's hill. F-2-5 moved out and secured the hill with no opposition. Once the artillery rounds started exploding, the enemy deserted the hill and moved back to catch us on the next one.

In a matter of a few days, Fox Company had lost about fifty percent of its men. I expected the company to go into reserve, but this never happened. We remained on the front lines with the men we had left.

Right after the skirmishes described above, we set up a defense on a mountain and were told to dig in and hold this position. So the next morning we remained in our position. Our outpost notified us that a lone Korean man in a white sheet was approaching our lines. We waved him on up the hill; he was carrying a set of barber tools. The CO let him cut hair for C rations. It had been a long time since any of us had had a haircut. I got a nice haircut for a can of corned beef hash. The barber left each evening and came back every morning. Every morning the word went out to watch for the barber and not to shoot him. Each morning he would come up from a different direction and cut hair all day. This went on for four days when one of our interpreters returned from the rear. He asked me where the barber had come from. I said that I didn't know, but that he had been cutting hair for about four days. The interpreter told me, "He is not South Korean, he is North Korean." I told the CO what the interpreter told me. The CO had the barber's box searched, and he found a map of our position in the bottom of the box. Every one of our mortars and machine gun emplacements were clearly marked on his map, including our command post. Each morning he had come up a different route to record whatever positions he could observe. He was a North Korean intelligence lieutenant and, I might add, a very good barber.

Baptism by Fire

F-2-5 was pulled back and in reserve until the company could get some replacements. It was down to a skeleton and needed replacements badly.

I utilized this break to get back to the battery for a couple of days. It was still below freezing and much too cold to wash and dry any clothing. I did the same thing I had done before: I heated some water on the diesel oil stove and, using a T-shirt for a washcloth, gave myself a good bath and had a couple of beers before returning to the front.

The evening I returned to Fox Company, I was summoned to the CP along with the gunnery sergeant. A lieutenant from naval intelligence had two Chines e prisoners that he wanted released in front of the Republic of Korea army without the knowledge of the ROK army. They chose the ROK army because the prisoners wouldn't come back if they knew they would be captured by the Koreans. This would be a very touchy mission. If the ROK caught them, it would mean death for the two prisoners, and some kind of retaliation, possibly torture, against us.

This was an experiment conducted by naval intelligence. Here were two Chinese boys, about sixteen or seventeen years old. They had been captured and taken to the rear. Instead of being put in the stockade, they had been given the royal treatment. They were given clean linen for their beds, and their meals consisted of steak, milk, ice cream, and cake. By releasing them, it was hoped they would induce others to desert the People's Republic of China army. I had a map of the area where the ROK army was located, with its position marked clearly on my map. I had no idea where the enemy lines were located, and I didn't have the slightest concept of the configuration of the enemy lines in the ROK sector.

Gunny Ski and I were to take a patrol out, release the two prisoners, and return the same day. I was not selected to go for artillery support but because I could read a map and get us there and back. I really didn't mind; I had been with Fox Company so long I felt I was one of the men. Our patrol would be made up of the new recruits in order to give them experience walking up and down the mountains.

We set out early the next morning with about twenty recruits. We were slowed down right from the beginning because the new men simply didn't have the legs for climbing the mountains. With a lot of breaks, we had spent over half the day climbing up and down the mountains and hills at a snail's pace, before we arrived at the base of the mountain occupied by the ROK army. We set up a plan whereby I would take the patrol up to the top of the mountain to the ROK command post, as if on a normal patrol. Ski was to take a few men, pull off to the side of the mountain on the way up, and release the prisoners. Each of the prisoners had been given two boxes of C ration before leaving that morning, so they wouldn't starve until they found their way back to some Chinese unit. Ski and I would then meet in a grove of trees that was quite visible to us. The grove was next to a river leading back to the company. We couldn't go back the way we had come; it would have taken us all night. This alternate route would be a shortcut and, according to my map, on much-welcomed flat terrain.

I was greeted by the CO at the ROK command post and was given a briefing on where the enemy lines were located. The Chinese were in the valley that Ski and I would have to cut across to reach the grove of trees for our rendezvous point. It was impossible to notify Gunny Ski that he was releasing the prisoners on the wrong side of the mountain. I didn't think that it would make much difference to the prisoners, but I was afraid Ski might get into a fire fight when he crossed over the mountain trail and back down into the valley. I had to carry through with our plans. It was too late to change now, even if it meant crossing enemy terrain.

I took the patrol back down the well-worn mountain trail to the base of the mountain. We left the base of the mountain on a secondary trail leading through a tall and massive bamboo thicket. I decided to give the recruits a break while we were hidden in this thicket. The recruits lit up and started smoking, laughing, and cutting up as if they were on a picnic outing. I was in front, sitting on a small bank beside the winding trail. I caught movement on the trail in front of me and brought my carbine up to fire. It was pointed at the chest of a Chinese soldier, and I was looking directly into the barrel of a Japanese .31-caliber rifle. We stared at each other; neither moved nor tried to speak. It was a standoff. The soldier had a bandage on his neck, and the men behind him had various wounds. These were the walking wounded, and they were crossing over the low point between the two mountains. After staring at each other for some time, I lifted my carbine and motioned him to pass. There were about a dozen; some had leg wounds and were limping. All of them gave me a puzzled look as they passed.

As they passed by the recruits, they kept turning and looking back at me. I believe they were expecting me to give the order to fire. Meanwhile the recruits were still talking and enjoying themselves, oblivious to anything else. When the last of the Chinese cleared the end of my patrol, I gave the order to move out quietly. I was afraid I was taking these men into a possible ambush; I could hear a big fire fight in progress just ahead, so it wouldn't have come as a big surprise. We followed the trail out of the bamboo thicket and down into the valley. The heated fire fight was still going on, and there was no way of knowing which side we were on. I couldn't recognize friend from foe at this distance.

I passed the word to the recruits that we might come under enemy fire at any time, and that they had better load and be prepared to fight. When we came out into a small clearing and the soldiers involved in the fire fight caught sight of us, they started running, ducking, and dodging as they scrambled for cover. I assumed this was the enemy and immediately gave the signal to double-time and run down into a ravine leading to the river. All of the recruits were right behind me. For some odd reason, none of them asked me for a break. We made it to the river to the large grove of trees. Gunny Ski and his men were already there, and I was very surprised to see the two prisoners still with them. Gunny Ski said he kicked their butts down the mountain, but they kept following him back like a dog. While our men waited for us they had taken the C rations away from the prisoners, divided them up, and ate supper. I told Ski the Chinese knew where we were, and we had better get out of there and fast.

The river was much too deep to cross, so we decided to follow a path leading back beside the river bank. On this path we had the river on our right and a tall sheer cliff on our left. The only thing between the river and the cliff was the path. This path had a slight depression or rut beside it, probably caused by water runoff. Gunny Ski was leading us single-file down this path when I heard the sound of two 76-mm howitzers fire. I yelled, "Hit the deck," as the fire came screaming in. I dove into a muddy rut just as the rounds exploded in the water beside the path, splashing water all over me and the men close by. I yelled for the men to run. We had only covered a short distance when two more rounds were fired. The guns were firing direct fire, but for some reason, they didn't have the correct angle to hit the cliff. They came very close to landing on the path. The guns continued to fire, and we continued to run this gauntlet of fire for over a mile before reaching cover out of the line of fire.

I worried about recruit casualties, but they were all accounted for and in good shape. These men were not the carefree troops that had set out on their first patrol that morning. They were no longer recruits. They had received their baptism of fire, and had been introduced to a serious war that played for keeps. These men had just danced their first boondocker ballet, and would finally realize the war was very near and very deadly.

The navy lieutenant was very disappointed in the outcome of our mission. Later he came by my foxhole and asked if I would be willing to try one more time and release the prisoners. I told him it was too dangerous; being saddled with a bunch of new recruits on such a touchy mission was out of the question. The lieutenant said I could pick the men I wanted if I would try one more time, because they had invested a lot of time and effort into this experiment. To take them back was to admit defeat and failure of the project. He said, "This time, take them out and lose them somewhere. I don't care how, but don't bring them back."

I took my three wiremen and set out before daybreak with the two prisoners. Each had been given a box of C rations. We wanted to get out and back as soon as possible. My men were good mountain climbers, so there were very few breaks. When we arrived at the base of the mountain, I pointed out the bamboo thicket and told my men to hide there and wait for me once they had released the prisoners. Halfway up the mountain my men cut off of the main path and went down the side of the mountain. I told them to fix bayonets and to really kick the prisoners' butts to make them leave; to let them know you really mean to kill them if they try to follow you back. I continued on up the mountain to the ROK command post. I arrived just before noon on a cold, windy day.

I was not greeted as before, but was immediately relieved of my carbine and taken to the commanding officer. The first thing he asked me was whether I was an officer. When I replied that I was a scout sergeant, he had me remove my parka. I then had to remove my clothing piece by piece down to my dirty long underwear. Each piece of clothing was thoroughly searched. They were very interested in my compass and map. I explained that I used the map and compass to direct artillery fire on the enemy. It was bitterly cold, and the wind whipping over the mountains had my teeth chattering. I was shaking from the chill, but I was afraid to complain or say anything. I stood in one place for over an hour before the CO returned. He asked if I had come up the mountain alone. I told him some of my men had come with me and were to meet me at the bottom of the mountain. He asked how many man had come with me. My plight was now critical; I dared not lie. I told the officer about the experiment and the purpose of the mission. I was afraid they would kill me for this, but if they caught me lying they would execute me and my men without any qualms.

I had been standing for almost two hours in my dirty underwear, and was almost frozen, when one of the Korean patrols returned to the CP. I noticed them showing the two boxes of C rations to the commanding officer. They had intercepted the two Chinese prisoners. I assumed they had been killed, but I dared not ask. Cold was the least of my worries at this point. My fate was being decided by a few command officers. The wait seemed endless. I was shaking all over. I couldn't tell if it was from the cold or from being scared. there seemed to be no separation between the two. At last the CO strutted over and stood before me. He looked at me for quite some time before speaking. He told me to put my clothes on. His men would escort me off the mountain. I was told to inform my commanding officer not to send any more patrols into his sector again, as they would not be welcome.

I was certain the reason they let me go was because they didn't capture my men. If they had caught them, I believe the Korean officer would have had us all killed. My men were hidden in the bamboo thicket griping about the long wait. They said they had fixed bayonets and had released the prisoners on the side of the mountain without any trouble.

It was getting late, so we took the shortcut. We never encountered any enemy on the way to the grove of trees, but we were on our toes when we started down the path beside the river. Either the enemy guns had moved, or the artillery didn't think four men a worthy target.

When I got back to Fox Company CP I asked where I could find the navy lieutenant. I was informed he had left at daybreak. He probably went back and reported his experiment as a big success. I never heard any more about this.

Uncanny flashbacks such as this continue to haunt my subconscious mind in a montage of nightmarish dreams, even to the present.

The Lost Patrol

The mountains of Korea are truly an artist's dream, if he or she wants to capture a magnificent snow scene. The majestic mountains are a dominant feature of Korea, but with a heavy blanket of snow, they are objects of breathtaking beauty. If you could observe this panoramic spectacle from a warm and safe haven, you undoubtedly would be enthralled by the sweeping beauty of this winter wonderland. However, if you are fighting a war, plodding along with a pair of worn-out boondockers practically frozen to your numb feet, you somehow fail to sense or absorb any of the pictorial assets of the country.

Our patrol departed at daylight and was led by one of the newer sergeants. My radio man and I went along to wreak havoc on the enemy should we encounter them on this patrol. About noon we had reached the limit of our patrol area, and were starting back, when it began to snow. Visibility dropped to zero within an hour, and with a strong driving wind, it had become a full-blown blizzard. I slipped and slid along with the other men, relying on the patrol leader to know the direction back to the front lines. He never said anything, but when it began to get dark I knew he was lost. I consulted the patrol leader; he had no idea where we were or in which direction we had been going. It was too late for me to identify any landmarks and get a fix on our position. We were on a mountain going in a westerly direction; that was all I knew. I thought we should be going south, but in a blinding snow storm you don't go plummeting down a mountainside in a fierce blizzard just to change directions. The sergeant elected to continue on the same course. I was worried that we might accidentally stumble into an enemy outpost. One thing was certain: we were still in no man's land. Everyone was chilled to the bone and ready to drop when we stopped around midnight. All the men crawled into their sleeping bags because that was the only protection from the cold, driven snow.

I asked the sergeant to set up a watch for the night. He wouldn't come out of his sleeping bag. He told me to pick some of the men and arrange a watch. I shook over a half-dozen men, but none would come out of their sleeping bags. I went to the most possible approach to our position and found an evergreen that gave me some protection. I crawled into my sleeping bag and leaned back against the tree to guard the trail. The snow was knee-deep; soon everything was solid white. I was still guarding the trail when I finally drifted off to sleep from exhaustion. Morning came much too soon, but the snow had stopped and the men began digging out, getting ready to move. The patrol leader continued on the same trail, and in the same direction that we had been going. About two hours into the march we came to a marine sound and flash-ranging unit. This post was set up five miles in front of the lines to locate and plot the enemy artillery and mortar fire by taking an azimuth on the flashes. Once I had my map oriented we were lost no more, and started back to the front lines and the company position. My feet were soaking wet and numb. I couldn't stop or take a break for fear my feet would freeze. It was late in the evening when we finally stumbled back into Fox Company line. It had been a very trying and torturous ordeal for all the men on this lost patrol.

After this incident, I never relied or trusted a patrol leader to know our correct coordinates on the map, even though many of the NCOs had a better understanding of map reading than most officers. The only way I could be sure of our position was to continually check and upgrade our position as we meandered through the maze of hills, ridges, valleys, and awesome mountains that were all around us, always looming to towering heights in the distance, and in front of us, as we continued our pursuit of the enemy to the north.

Punch Bowl

After many small and large encounters with the enemy, we finally arrived
on the hills overlooking the punch bowl. It was called the punch bowl
because of the wide, circular expanse of low land surrounded by miles
of hills and mountains. On the evening of our arrival at the punch bowl,
the F.O. officer was all set to fire night defensive fires. This came first, but
on this evening our new radio man had not arrived. I backtracked down
the mountain and found him about halfway up and still climbing. He was
a new man and was handicapped by the weight of the radio, and by legs
that were unaccustomed to climbing steep mountains. It was very late
when we got the radio and finished firing our night defensive fires. The
radio was a heavy load, and we had to get a stronger man to carry the
weight up and down mountains. We requested and got a stronger man
who was capable of keeping up while carrying the heavy and cumbersome radio.

We moved to various positions along the rim of the punch bowl,
always sending out probing patrols. The enemy seemed to be holding the
north part of the punch bowl, and would offer some opposition every time
we encroached on their territory.

In one position, we dug in and continued to make daily patrols to the
north rim of the punch bowl. This was about five miles in front of our
position. We started to alternate the daily patrols with Dog and Easy
Companies. In this way you only had a patrol every third day. We always
patrolled out to a large hill, and this would be the limit of our patrol in no
man's land.

On one of these patrols we found a squad of ROK soldiers manning
an outpost on this hill. Directly to the front was a larger mountain which
was occupied and controlled by the North Koreans. This outpost had

foxholes dug about twenty yards apart, and they were being hit every night. None of these soldiers had automatic weapons. I decided to give the enemy a big surprise. I traded my M-2 automatic carbine for a Word War II M-1 semi-automatic carbine. The lieutenant in charge set up a machine gun on the trail from the big mountain leading up to that outpost. They now had two automatic weapons to greet the enemy with when they charged up the hill on their nightly raid.

On our next patrol we found the outpost abandoned. The soldiers and our weapons were gone for good. They left evidence of what a difference automatic weapons can make. Over twenty enemy bodies were scattered along the trail approaching that outpost. The weapons had served their purpose.

On another one of these patrols we came under artillery fire from the two infernal 76-mm howitzers. The only protection we had was to dive into the old foxholes left by the men who had previously manned the outpost. The artillery was being fired over the shoulder of the big mountain, and I was unable to determine its location. The gun emplacements were hidden behind the huge mountain. It seemed these two artillery howitzers were determined to get me. They started plaguing me from the first time I set foot on the front lines. They generally fired on our company or patrols, but being a scout sergeant, I began to take it personally. I was determined to get them if I possibly could.

On our next patrol we arrived on the hill in the late afternoon. I asked the lieutenant to hold the patrol while I walked down the ridge line off to our right to try to spot the artillery guns that were hidden behind the mountain. I wanted to call a fire mission and destroy the guns. My radio man and I went over a mile along this ridge line until we came to a pock-marked hill. It's hard to guess how many artillery rounds had exploded on this hill, but I figured I was in the right position. I laid down and scanned the terrain behind the mountain with my field glasses. I could not detect the two howitzers.

I was about to give up when I spotted four guns poking muzzles out from under the camouflage netting. It looked like one of our artillery batteries set up and ready to fire. These were not 76-mm howitzers, but 105-mm howitzers. They were located on a high plateau and could easily fire across the shoulder of the mountain. I tried to adjust our artillery on the guns, but they were out of our range and I had to give it up. The patrol leader was mad because we took so long. It was after dark when we got back to the company. I called Capt. King at the battalion liaison to try to get some kind of artillery fire on these guns, but I was told they were beyond the capabilities of our guns. I called, asking for an air strike on these coordinates. Capt. King from battalion called me back and said the request was denied because no one had reported receiving any 105-mm artillery fire. They were telling me I didn't know what I was talking about. It's easy for someone in the rear to make decisions like this, when they are not facing the guns. It was very late, and I gave it up for the night. The following day Dog Company took its turn on patrol. As soon as it arrived on the hill, Dog Company was hit with a surprise barrage of over seventy rounds of 105-mm howitzer ammunition. Dog Company lost a number of men in this barrage. One of the dead was a black marine. He was a likeable marine, and the first one I had ever seen in a combat unit. Many times when our paths crossed on the trails we would stop and talk. After Dog Company was hit, everyone called me, wanting to know the coordinates of the guns. I gave it to them gladly. I never knew what action was taken, but we never received any more 105-mm fire. For the men in Dog Company it came one day too late.

Fox Company had its own mail orderly. The mail for the F.O. team was generally brought up the mountain by one of our wiremen. Leaving the mountain, and going down and through the winding valley was an all-day chore, and at times you came under sniper fire. Often our wiremen would delay bringing up the mail because of the snipers. One day I decided to go back down the mountain and back to our battalion liaison and pick up our mail. I wasn't that scared of snipers. At times the snipers had come very close, but this day I had no problem going down the mountain and through the winding valley. No one fired at me as far as I knew. Sniper fire never worried me in this kind of terrain. A sniper firing from the side of a hill or mountain would have to be an expert sniper, or just plain lucky to do any damage. The range of an elevation can be very deceiving when firing from a higher elevation to a lower one. At times I've had sniper fire come very close. Any marine who has ever worked in the butts on a rifle range can tell you when a shot is very close. It generally makes a loud cracking noise. When this happens the sniper is very close to being on target, and if you start running, he will snap off as many rounds as he can to try and hit you by using the shotgun effect. The best defense is to ignore the fire and to continue walking at the same pace, so the sniper won't know he is close to his target. Most likely he will readjust his sights and then miss you by a wide margin.

I picked up the mail for the F.O. team and was given a manila envelope addressed to the commanding officer of Fox Company. It had a large printed stamp on the front that read, "Top Secret." I was surprised the envelope was entrusted to me, but I was told that I was the only one who could get this envelope to the company in time.

In time for what? It was a long trip back up the valley, and then the mountain, carrying such a top secret envelope. My mind was in turmoil, trying to guess what was in this envelope. What did time have to do with anything? I was still guessing when I delivered the envelope to the skipper.

The envelope was to be opened at exactly eight o'clock that evening. Everyone waited anxiously and speculated on what the envelope contained. What would be revealed on the grand opening? The envelope was opened right on time. It had to do with military scrip, the paper money used by the military. A new and different colored kind of scrip would replace the old scrip. This scrip would be legal tender at eight the next morning. Anyone with the old scrip would have to turn it in before then, so it could be replaced by the new. At eight the next morning the old scrip was just like confederate money. It would be worthless. This whole secret operation bypassed the civilian population to scuttle the black market operators. It was a good plan, and probably saved our government a bundle. Keeping it top secret insured its success; it couldn't fail.

A new lieutenant was sent up as my F.O. officer. In the two weeks he was with me, he never fired a mission or made a patrol. He appeared to be shell-shocked. He stayed close to his foxhole, and would dive in at the faintest sound of gunfire. He gave me his liquor ration in exchange for me making all of his patrols. I didn't drink liquor, but I took it and gave it to the men in Fox Company. The lieutenant was so scared and edgy that he made me nervous.

One day when we were not on patrol, I took a day off to go back to the battery, pick up the mail, and try to get some clean clothing. The men were now calling me Jungle Jim, and it wasn't because I needed a haircut.

Capt. Les Procter greeted me on my return to the battery, and asked how the new lieutenant was doing on patrol. The captain said he had sent him up to get some experience as a forward observer. I told the captain the truth: the lieutenant had never fired the guns or made one patrol, and didn't seem interested in learning how. The captain told me to remain back with the battery for awhile. In this way the lieutenant had no alternative but to make the patrols. I was put in charge of battery security. I had time and was able to wash my clothing. I felt like a new person.

The captain said he was willing to recommend me for a battlefield commission if I was interested. I had to turn it down. I still hadn't made up my mind if I was going to stay in the Marine Corps or if I was going to try civilian life again. I asked the captain to send me back to Fox Company. The captain said he needed a battalion liaison man to aid the Korean Marine Corps' forward observers. This was my next assignment. I never went back to Fox Company.

Korean Marine Corps

I spent over a week maintaining security around the perimeter of B-1-11. I didn't feel at ease being in the rear. I did enjoy the warmth of sleeping in a tent. I thought sleeping on a cot invited trouble. I wanted to get back to the front lines and a good, reliable foxhole. I did appreciate this break as it gave me time to bathe and wash and dry my clothing.

I was also presented with the Silver Star while trying to stand at attention in ankle-deep mud. This decoration was for the bayonet charge while I was with F-2-5.

Instead of returning to F-2-5, I was ordered to a battalion of the Korean Marine Corps regiment. I was to maintain communications between the Korean forward observers and the rear liaison station.

The Korean Marine Corps was a regiment of Koreans that had been trained by the marines. The Korean Marine Corps had already established itself as a good fighting unit before I joined them. The company weapons of a regular marine unit were .30-caliber light machine guns and 60-mm mortars. For some reason the Korean Marine Corps carried .50-caliber machine guns, 80-mm mortars, and flamethrowers as their company weapons.

I joined the Korean Marine Corps carrying an old switchboard and accompanied by an interpreter who had been assigned to me. I planned to set up land lines between the forward observers, my switchboard, and the battalion liaison back down in the valley behind our mountain. I was to receive the fire missions from the Korean Marine Corps forward observers, and relay them back to the battalion liaison. They, in turn, would relay this by radio back to the Eleventh Marine Artillery.

Setting this up proved to be quite a problem. My wiremen laid the wire, but the switchboard wouldn't work. The battalion said they didn't have another switchboard but that they would requisition one. I was told that even if they found one, it would be two or three weeks before I received it. I tore into the old switchboard and rewired part of it, using and installing a sound-powered phone. It didn't work right, but it was operational. My next surprise—my interpreter had learned his English in the marine mess while scrubbing pots and pans. He had a kindergarten vocabulary and many of his words were not suitable for useful functions. Only another marine could fully understand him.

Only two Americans were on this snowy mountain: myself as artillery advisor, and one infantry lieutenant as the infantry advisor. He had his own interpreter, who wasn't much better than mine. The two would-be interpreters stayed close to their bunker, and the only time I saw them was at briefings on different objectives.

I felt like a recluse being isolated on an icy mountain with no one to talk to.

This whole operation was like a Chinese fire drill, but I got the job done. I can't relate all the fire missions I relayed, but to show what I was up against, I will give a brief description of one of the many fire missions I handled while with the Korean Marine Corps. Sgt. Ode was a scout observer who was located on an outpost about a mile in front of our position. Late one evening he requested a fire mission. I could hear the small arms fire and knew he needed the artillery fire immediately. My interpreter was always visiting with his Korean friends, and I didn't have time to run him down. I was able to get all the information I needed to get artillery fire on his coordinates, all except the nature of the target. Battalion wanted to know the nature of the target before firing. I explain as best I could to Sgt. Ode. What was he shooting at? He replied in a few English words, "Firing on enemy foxhole." I told battalion to go ahead and fire the mission, because the sergeant's outpost was being attacked. Battalion fired the mission, and Sgt. Ode requested a repeat fire for effect. After this he gave a cease-fire end of mission. The battalion requested a surveillance report of the damage. I tried to explain to Sgt. Ode that the battalion needed to know what he destroyed with the artillery fire. He said he understood and politely said, "Cease-fire, end of mission, destroyed one enemy foxhole." His English was limited, but later I learned he had the enemy stacked up knee deep, but didn't know how to report this. He was a good scout sergeant.

My phone lines back to battalion went out one morning, and I radioed back to battalion to get the wiremen out to repair the trouble. It was late in the evening when they arrived on the mountain. They found the break in the lines in front of Maj. Haun's bunker. The major had caused the break by chinning himself every morning on the communication lines running through the tree limbs just above his head. My wiremen repaired it and made the long journey back down the mountain.

About two weeks after this interruption, we had another outage. Instead of having my wiremen come all the way up the mountain, I decided to repair it myself. I started in front of the major's bunker, but he was not the culprit this time. I continued to troubleshoot the wire for over two miles along a ridge line exposed to the front. How much exposed I wasn't sure. I kept pulling the wire up out of the snow until I finally found the break. Someone had built a fire over the wires, burning one of the wires in two. I was wearing a parka cap and a parka coat, but I was still wearing my old boondockers and a regular pair of gloves. I didn't have any cold weather gloves or any tools to repair the break. I used my marine corps K-bar to peel the insulation from the wires. My feet were numb, and my hands became numb while trying to repair the wire.

Someone started firing at me from another ridge line that ran parallel to the one I was on. It wouldn't have surprised me to find it was the Korean Marine Corps doing the firing. I had enough slack to pull the wires over to the reverse slope for protection until I could repair the break. I had stripped away about six inches of insulation from both ends of the broken wires before my fingers went completely numb. I managed to make a large square knot. In an effort to tighten the knot, and in trying to pull it tight, I put one end of the wire between the palms of my hands and the other end between my teeth. I had just started to tighten the knot when someone cranked the field phone. I saw a massive display of fireworks before I was able to drop the wire. I kept feeling the wire and waiting to see if anyone else would crank the phone. After waiting a bit, I tried the same thing, and again someone cranked the phone. Being caught again with the wire in my mouth, I again witnessed the varied spectrum of colors and the bright display of lights. I quit. I was through. I had tightened the knot some, not like it should have been, but it would do for now. I didn't dare show myself back on the ridge line.

I suspected someone was waiting to give me a big surprise, so I trudged through the snow on the reverse slope for about a half mile before surfacing to the top. My hands and feet were numb, but I could not stop. No one knew I was out here, except my interpreter, and he would be the last person to leave the mountain in a search for me. I was just able to make it back to my foxhole. I was very lucky. The major's orderly took me into his bunker to get me warm by his small fire. He cooked me some rice and kimchi. This was a welcome feast and totally unexpected. After this the orderly and I became friends. I would give him my C ration, and he would cook me two hot meals a day, mostly rice and kimchi or some kind of soup poured over the rice. The orderly cooked the same thing for the major. He knew English better than my interpreter, and he also knew just about everything that transpired on this mountain. He was the source of my information as long as I was with the Korean Marine Corps.

The Korean Marine Corps began making night patrols. I had never made a night patrol in Korea, so I was reluctant to send a marine F.O. team on this kind of patrol with the Koreans, but I was ordered to send them out to give the Korean Marine Corps artillery support if they encountered any opposition.

My foxhole was on the reverse side of the mountain, away from the enemy front lines. With the patrol out in front, I would have to take my radio up on top of the mountain to communicate with them. I set the radio up in freezing weather. It was well below zero, and the bone-chilling wind whipping across the mountain was torture. Fine snow was ricocheting all over the mountain top.

It was a very cold wait. I started crushing ice with my toes in my boon-dockers. This ice had formed from the moisture and perspiration that had accumulated inside my shoes. I had to do something or suffer frostbite. I was all by myself. I took my sleeping bag up to the peak and pulled it up to my hips. This would help stimulate the circulation in my feet and keep them from freezing. I had to maintain this station; without it the F.O. team would be useless.

It was over two hours before I received a faint and broken signal from the F.O. team. It was unreadable, but another F.O. team somewhere in those mountains relayed the message to me. They were lost. The Korean Marine Corps went so fast that the F.O. team fell behind and had taken the wrong path. I took their coordinates back to my foxhole. The wind was blowing so hard they could not read their map. They gave me the coordinates of their position just before they moved out into a valley. This valley was all in red on my map, indicating a dangerous mine field. I sent the message about the mine field to the F.O.s and asked them to try to backtrack, being careful to remain on the trail. I requested the relay F.O. radio to stay on the air with me until I had them back. They said they were freezing and wanted to secure the radio, but they would stay with me as long as they could. The Korean Marine Corps patrol returned, bringing some killed and wounded back on stretchers. I briefed the patrol leader on the lost F.O. team. He immediately sent out another patrol. During all of this, the relay station pleaded for permission to secure his radio. I knew they were cold, and I hoped none of them were wearing boondockers. It was over an hour later when I received the message that the Korean Marine Corps patrol had found the men, and that they were on their way back. I thanked the relay station for its help. They secured their radio without one word. I didn't know this man, but I remembered his handle. Thanks again to Oboe Two.

The flaps of my parka cap were frozen solid from the moisture from my breath. It was after three in the morning when I secured my radio and was finally able to crawl into my foxhole and sleeping bag to thaw out and get some sleep.

I never sent another F.O. team out with the Korean Marine Corps and I was never ordered to do so after this incident.

I had been on the mountain for about three weeks when battalion sent me another switchboard. I hooked it up, and it worked normally. I sent the old one back down the mountain. The next morning a communication sergeant called me and wanted to know how I had operated the switchboard without an audio transformer. I couldn't explain how I got it to work. It was by trial and error, wiring all the lines through a sound-powered phone. It worked, except anyone else could listen by picking up the phone. This didn't matter, as there were never any big secrets being relayed over the switchboard.

Stealing in the Korean Marine Corps was punishable by death. The executions were carried out promptly and on the spot. There was never a court martial nor appeal. It didn't seem to matter; these men stole anything they could lay their hands on, even your steel helmet if you left it outside your foxhole. On one occasion, a Korean Marine Corps soldier, whose foxhole was located close to mine, was searched. They found a wallet that had been stolen from another soldier. This man was taken to the commanding officer's bunker for interrogation. All day long you could hear the screams coming from the bunker. Late in the evening two soldiers helped him back to his foxhole. My interpreter told me they had tortured him by striking him on the elbows, knees, and ankles with a telephone or radio receiver. I asked Duffy, my interpreter, if the man had admitted to stealing the wallet. He said, "No, if he had confessed to stealing the wallet, they would have shot him." The man maintained someone had planted the wallet in his foxhole to get him in trouble. It was over a week before the man could walk again.

It was around Christmas 1951 when I was relieved of my duties with the Korean Marine Corps. I stopped in the valley and thanked Lt. Francis Gore and Lt. Thomas for all their help and assistance while they were in charge of the battalion liaison. I was glad to be leaving the Korean Marine Corps. They were good combat troopers, but I missed being with the men in F-2-5. I hoped I would be able to rejoin them.

B-1-11

Returning to B-1-11 was like joining a new unit. I only knew a few of those men, and they were the forward observation teams who were serving with the Fifth Marines, Dog and Easy Companies. It didn't really matter. I had been in Korea for a little over twelve months, and I never really cared to generate a close, friendly relationship with anyone. We were always losing men, and a close relationship only created more intense traumatic circumstances.

When I returned to B-1-11, I hoped to go back to Fox Company, Fifth Marines, or maybe R&R to some liberty town. I was surprised when the CO advised me that I had completed my tour of duty, and I would remain with the battery until transportation was available to the States. This was news to me. I didn't know there was a limit on a tour of duty in Korea. I thought I would remain to the very end.

While I waited, I was put in charge of battery security, and one of my duties was to keep the surrounding area patrolled daily for any sign of enemy activity.

Sgt. Ernie—his last name is immaterial at this time—was the scout sergeant who was stranded in the mine field while with the Korean Marine Corps night patrol. He was also waiting to leave for the States with one group the following day. I assigned him as patrol leader for one of the routine patrols. He came to my tent and asked me if he could be excused from his patrol. He said he had a bad feeling that he would be killed and would never make it home. This was surprising because he had been on the front lines and had been subjected to enemy fire. I couldn't believe he was really serious, asking me to do this, but I went ahead and excused him. I assigned another sergeant to the patrol. I remembered my old boot camp buddy, Cpl. Campo, who had a premonition that he would be killed on Okinawa. I understood the sergeant's feelings. The closer you come to leaving a combat zone, the more apprehensive you become. Most all short-timers, as we called them, were afraid to take any unnecessary chances. The patrol went out and returned without any mishaps.

While in B battery, I found a recent copy of the casualty list of F-2-5. I recognized a number of names on the list. The company had been hit hard, and it left me wondering if the F.O. officer or scout sergeant had fired their night defensive fires. This fire could easily compromise an enemy's attack, giving the defending company or battalion an edge in defending their positions. I felt left out and was sorry not to have been with F-2-5 during its last encounter with the enemy, even though while I was with Fox Company, I didn't get to join in every skirmish. At times the company would be poised to assault the enemy, and I would be back on a hill behind them, firing artillery on top of the hill to soften the enemy's stronghold, before the infantry company jumped off in the final assault. I could never recall all the fire fights and the many skirmishes that took place while I was on the front with F-2-5, so I've only related a few that have remained the most vivid in my mind.

I finally received a pair of shoepacs, the cold weather boots. I discarded my old boondockers. They were worn out and about as thin as an old worn-out wallet. Those boondockers had given me many miles of good service.

Sgt. Ernie and his group had departed over a week before, and I was now in the same position he had been in. I would be in the next group leaving the following morning.

The night before leaving, Capt. Les Procter, our commanding officer, called all the men who were leaving into his tent. We sat around in a semicircle on the floor, or the tent deck. The captain said, "I know that most of you don't have any really good memories to take home with you, so I called this meeting to try and give all of you at least one good memory before you leave." After finishing his short speech, he passed around a bottle of Scotch whiskey, and we had a drink and sang some songs. We then passed the bottle around for the last time. The captain was right. It was a memorable evening. I have never forgotten this farewell party. We had shared a good part of our lives together. The curtain was drawn. This would be the last time I would ever see those men.

We departed early the following morning, and had to walk to the battalion CP to board a truck. My feet were freezing inside my shoepacs, and the shoepacs made blisters on my feet. It was too late now; my boondockers were long gone. I was just thankful I was going home and not back to the front lines.

I had been in Korea for just over a year and never had a liberty of any kind. R&R—rest and relaxation—is something you hear about, but it's always for someone else and never for you. It's just as well because I never had a good bath either while in Korea. The men in F-2-5 probably still refer to me as Jungle Jim.

My feet were numb when we finally boarded ship. The ship was like heaven compared to what we had been accustomed to. Once we raised anchor to get underway, I went topside to watch the land of Korea fade away. When I had first set foot in this land, my first thoughts were that I would never leave it alive. That seemed ages ago. It was all over, and I had survived. I had the honor of serving and surviving with an elite group of fighting men under the most adverse circumstances imaginable. I have memories that are etched indelibly into my conscious and subconscious mind. I hoped this would be the last war that would offer a challenge to my survival. As the last traces of land disappeared over the horizon, I went below, took a shower, crawled into a bunk, and fell sound asleep. War is an exhausting experience.

Homeward Bound

Back in 1945 I had departed Japan from the small Japanese town of Sasebo. Our ship had docked in Kobe, a much larger town. I thought this town would be similar to Sasebo, but it was a large metropolitan town with tall buildings and traffic jams. Our purpose in docking was to retrieve our sea bags that had been in storage while we were in Korea. I enjoyed the liberty in Kobe. I ordered a large bottle of Osaka beer in the first bar I visited, and while drinking the beer I observed the friendly and pretty girls. Two shore patrolmen walked in and said, "Sergeant, you are in an out-of-bounds area; you'll have to leave." I did the same thing in another bar, and again the same two shore patrolmen told me to leave. I went over two streets, ordered another bottle of beer, and had a really pretty Japanese girl sitting on my lap, when the two shore patrolmen entered. They said, "Sergeant, this is the third time. If we catch you in an out-of-bounds area one more time, we'll escort you back to your ship." I told them I had looked all over, and I couldn't find an out-of-bounds sign. They politely informed me that there were no signs; any time you found a place that had beer and women in the same building you could be sure it was out-of-bounds. I gave up trying to relax and drinking a beer. I spent most of my time looking around the various stores and shops.

My uniform smelled musty after being in the sea bag for over a year, but it wasn't too noticeable combined with the smell of the city and the bay. The town reeked with a polluted and foul odor, like a salt water marsh combined with the city's polluted air.

The new shoepacs were clumsy and never fit my feet properly. I was really happy to discover another pair of boondockers in my sea bag.

I met Tankersley, my friend from Tent Camp Two. He had survived Korea but gave me some bad news of other men who had been in our training platoon who wouldn't be going home. Tankersley had a small chess board, so we passed many hours playing the game.

I made friends aboard ship with a navy chief. He was retiring once the ship reached San Diego. This was his final tour of duty. The ship's crew gave him a big party with a big retirement cake. Only a few marines were invited to the party, so I thought it an honor when I was invited. I congratulated the chief on his retirement. He thanked me and then told me a humorous story that had recently taken place in Tientsin, China. He told me that ever since he had been in the navy he had heard the marines were first to land, first this and that. He went on to say, "I never believed this crap, but after my tour of duty in Tientsin, I'm convinced the marines have been everywhere and may be the first to land, or whatever." He said, "I met a beautiful Chinese girl in Tientsin, and you could tell just by looking at her that she radiated purity and was probably a virgin. I decided since I was retiring, I would marry her and take her back to the States with me. I had a couple of dates, wining and dining her. She was always dressed in clean and expensive kimonos. After proposing, and a lot of talking, I finally convinced her to go with me to a hotel. This was to be the happiest day of my life. I was anxious and quickly got undressed. The pretty little Chinese girl took off her clean kimono, and then proceeded to peel off over ten green Marine Corps skivvy drawers, each one with a different marine's name stenciled on the front. I'm now a firm believer that the marines are always the first to land." This was a good party, and a welcome relief from the ship's routine.

We made a brief stop in San Francisco, and Tankersley and I got up to go ashore. Many of the men had called home, so there was a large crowd on the docks to greet us when we docked in San Diego. Once ashore, we were billeted in the barracks across from the marine boot camp. I visited the base PX, and was enjoying a big milkshake, when I met one of the F.O. radio men who had been in Korea. After his return from Korea, he had been assigned to the communications school. He asked if I had heard the news about Sgt. Ernie. I said, "No, I just arrived yesterday." The man said that Sgt. Ernie and some other marines were on their way home when they were involved in a car accident. Sgt. Ernie was killed in the crash. This came as a shock to me. His premonition was right. He had made it out of Korea all right, but he never made it home. What a shame.

It doesn't matter where you are—in a combat zone or on some scenic road—when the Grim Reaper beckons, you're gone. This has always been beyond my comprehension.

Tankersley and I learned that it would take thirty days to be processed for discharge, but if you extended your enlistment, you could leave immediately. We made a hasty trip to the office and extended our enlistment.

I loaned Tankersley forty dollars so he would have enough money for plane fare. I couldn't refuse. I remembered the time he befriended me by loaning me twenty dollars when I didn't even know his name.

I had my share of close calls. I wasn't about to give the Grim Reaper an edge. Instead of flying, I took a Greyhound bus and spent a long and dirty three days' journey to San Antonio, Texas.

Camp Lejeune,
North Carolina, 1952

I had a happy reunion with my wife and a memorable stay in San Antonio. I cut it short because I wanted to visit a few days with my folks in Dallas. Dallas had grown so big, I was afraid to drive on some of the freeways.

On my arrival home I found a forty-dollar money order waiting, and a nice thank-you note from Tankersley's wife. My orders to report to Camp Lejeune were also waiting. It was a blanket order, and I noticed Tankersley was assigned to a camp on the West Coast. I wished I could have maintained this friendship, but the distance was too great. We lost contact, and I never saw him again.

It was really good to be home, if only for a short while. I had a nice visit and was really surprised to see how big my little brothers and sisters had grown. I still remembered them as babies.

On arriving in Camp Lejeune, the first thing I noticed was that the camp is on flat terrain. This was quite a contrast, and a lot better than Camp Pendleton, with the barracks located on numerous hilltops. I had no trouble locating the headquarters of the Tenth Marines and was given a layout of the regimental barracks. I reported to the Third Battalion, Tenth Marines. On arriving at battalion, I was greeted by Sgt. Maj. William S. Prosser. Sgt. Prosser had been my section chief during the Pacific campaigns. It was good to greet an old combat trooper and see a familiar face. This was the first I learned that Sgt. Prosser had remained in the Marine Corps after World War II. When it came to being a 100 percent marine, Sgt. Prosser was always a 110 percent marine. Later I chanced to meet two other men from the old F-2-10, 75-mm howitzer battery. One was Sgt. Wyatt and the other Sgt. Bill Moss. I also learned that Sgt. Schott had stayed in the marines and was teaching artillery on the West Coast. He had been my section chief when we set the gun up on top of the coral cliffs of Tinian during the battle for the Marianas. I was never able to make contact with him.

I was assigned as section chief in H-3-10. I had committed myself to being a career marine, so I decided to settle down and apply myself to learning all I could to further this career.

When it came to studying and learning a subject, I was like a sponge, able to absorb just about any subject. During the first year I completed and received my GED diploma and then completed six schools from the Marine Corps Schools in Quantico, Virginia. The first course was Marine Corps History, followed by Basic Officers' Indoctrination Course, Marine Corps Administration One and Two, Aerial Map and Photograph Reading, and the last was Military Justice under the Uniform Code.

Soon after reporting to Camp Lejeune, I was assigned to six weeks of radiobiological warfare training in Frenchman's Flat, Nevada. This was called "Desert Rock Two." Up until this time no one had ever been closer than ten miles from the blast of an atomic bomb. This would be the first time anyone would be located only four miles from the blast site at ground zero. This blast would culminate our six weeks of training. We had an early reveille and chow and were driven out to the test site, where we were each issued a radiation badge to measure the amount of radiation each man received during this test. We had a long, cold wait until countdown. We were instructed to stand up once the bomb was detonated.

The countdown started, and suddenly there was a brilliant flash of white light. I stood up, but I couldn't look directly at the blast. There was a ball of fire so hot that even at this distance it seemed as if you were standing in front of a raging inferno. It slowly cooled to a fiery glow. Only then could you look directly at the orange ball and the mushroom cloud rising over the desert floor. The sound of the blast wasn't acute but rather a continuous roar. I watched with my mouth open. When the wind and shock waves that expanded out from the blast at ground zero reached our position, I got a mouthful of sand, grass, and dirt, generated by the speed of the wind as it whistled past your ears. I turned away from the blast just in time to get another mouthful of the debris, as the wind rushed back in to fill the vacuum created by the explosion.

We boarded trucks and crossed over ground zero exactly one hour after the atomic blast. We walked around and across ground zero, observing the damage inflicted on the dummies and equipment that had been placed on this site. This concluded our six weeks' training. Our badges were picked up before our return to camp. No one was ever told what doses of radiation we had received during this maneuver.

Once back in Camp Lejeune, I decided if I was going to remain in the

Marine Corps that I would apply for a commission. I submitted my application for the platoon leader's course in Quantico, Virginia. This school was open to all NCOs with over seven years' service. On completing this course you were commissioned a second lieutenant. The NCO had to have a high level IQ and appear in front of a screening board of officers. I had all of the qualifications. I didn't know what to expect when I appeared in front of the five officers on the screening board. It was like a third degree; they covered just about everything they could come up with. I was recommended with enthusiasm. This was the highest of the five possible recommendations. Gunny Ski from the Fifth Marines had just completed this course and had received his bars as second lieutenant. I was waiting to attend the next class when the Marine Corps received a cut in appropriations. The platoon leader's course was the first thing on the chopping block. With this door closed, the only thing I could see down the line was the possible rank of sergeant major. I didn't care for office duty, so the only other option was to strive for the rank of master gunnery sergeant. This was a field rank, but this rank was rumored to be one being dropped from the Marine Corps. My future in the marines didn't look too promising. With possible promotions blocked, and no war to fight, it seemed useless to stay in the Marine Corps. I was already becoming bored and extremely depressed at times. The only positive things in my life were my wife and new son who had recently been born in the Camp Lejeune naval hospital.

Trying to maintain my family and still fulfill my duties to the Marine Corps proved to be a full-time job. I learned to cope by talking to and observing other married marines, or brown-baggers. This was what the marines called married men. We had some rough times and some happy times. It wasn't easy for my wife, but for a while we settled down to a stable and normal married life. However, I soon learned there was no such thing as a normal married life for a member of the fleet marine force.

I was a staff sergeant, but I was assigned as acting gunnery sergeant for G-3-10 until my promotion to gunnery sergeant came through. Promotions in artillery were extremely slow, so it would be some time before my name reached the top of the promotion list in the Marine Corps headquarters.

The second morning as gunnery sergeant of G-3-10, I faced my first big challenging assignment. Standing in front of me were 151 burr-headed recruits, fresh out of boot camp. None of these men had ever seen a 105-mm howitzer. I had only three months to train and shape them into a synchronized and smooth-operating battery before participating in the scheduled training exercises in Fort Bragg. In Fort Bragg we would be in competition with the other units and would be graded on our performance.

The executive officer and I set up a strict training program that would keep the men busy. This included one overnight training exercise each week. I was very fortunate to have received a group of men eager to learn. They reminded me of the naval academy midshipmen.

After an expenditure of a lot of time and hard work, coupled with the fine assistance of my section chiefs and NCOs, we developed an artillery battery that I would have been satisfied with in any combat situation. I enjoyed training these men. It helped to overcome the extreme boredom that seemed to plague me. I didn't enjoy playing cowboys and Indians and not having anyone shooting back.

Three months after starting the training, our battery participated in the Fort Bragg maneuvers. This corresponded to a graduation performance, and it came off like clockwork. Our battery took top honors in the exercises. We had trained the men well. I was proud of them. They operated like a team and displayed excellent teamwork. The executive officer was nice enough to take charge of the battery and let me go up with the foreword observers and fire a few missions. This was the first time I had fired a mission since I had left Korea. It really seemed odd to waste a lot of ammunition without trying to destroy or kill someone.

Vieques, Puerto Rico and the Mediterranean Cruise

Vieques, Puerto Rico is a small island situated a short distance across the water from Roosevelt Roads. The naval station is located on the main island. Looking toward the big island from the small town of Isabella Segunda, you can see the naval station, and looking above the naval station you can observe the lush tropical growth of the rain forest extending all the way to the top of El Yunque Mountain.

We would be on the island of Vieques for three months undergoing some extensive training in combined navy and marine amphibious maneuvers. We would be practicing amphibious landings of field artillery in assault situations. The training we had given these men for assault situations was quite evident in their fine performance in the various maneuvers. It made my job a little easier, and it was a pleasure to see the battery operate smoothly as a team.

We were scheduled for our final and graded maneuver. We would be in competition with the other batteries.

The executive officer and I had always been agreeable as to the training and operation of the battery. At times we had some small differences, but we could always talk them out and come to a reasonable compromise or solution. This would be the first exception. The executive officer had made a reconnaissance patrol of the area and had selected our gun positions and the site where we would park the DUKW (duck with crane or A-frame to unload the howitzers). I advised the lieutenant that we were near a rain forest, and that this was the tropics. To unload the guns anywhere but on the crushed rock of the road was taking a chance of being bogged down. This terrain had a crust, and once the crust

162

was broken you were stuck in the mud. The executive officer became very upset, and told me to unload the guns on the coordinates he had selected.

We landed in a slow, drizzling rain on the morning of the graded maneuvers. The landing came off perfectly. I moved the battery inland and set up the A-frame on the selected position to unload. The DUKW with the first gun section drove under the A-frame on the rear of the A-frame DUKW to unload. The A-frame connected the winch to the cable slings on the gun and raised the gun up, so this DUKW could pull out from under the gun. The A-frame DUKW with the added weight of the gun broke the crust and subsequently mired in the mud. When it sank down the gun was straddled over the gunnels of the other DUKW. We spent the biggest part of the day getting the gun off the DUKW and onto the ground. We got the A-frame out of the mud and back to the coral road. We were then able to unload the other guns and get them into position and ready to fire. Our battery came in last on this maneuver. I was so disgusted I never asked what kind of a grade we had received.

This island was very small, and the only town was Isabella Segunda. The only thing you could get on liberty were a few beers and a steak dinner. This had to be eaten outside in the open. The flies were no problem but the mosquitoes were a nuisance. Most of the people in this town were black and even appeared to be hostile at times.

I had some Puerto Rican marines in our battery They were given liberty to the big island. I received a weekend pass by saying I would be staying with one of the men who lived in old San Juan. He was from a big family, and they didn't have the room to put me up for the night. All I wanted to do was to visit old San Juan then go back to the liberty boat docked at Roosevelt Roads. I enjoyed the liberty and looking the town over. I had a few beers in some of the bars and was surprised not to see women in any of the bars. I caught a bus from the circle in San Juan going to the town of Santurce. Here I would be able to catch a bus to the San Juan International Airport. Once in Santurce, I had over an hour to wait. I walked a couple of blocks to a place called the Zombie bar. I bought myself a beer and noted the only other customer in the place wasn't drinking, so I ordered him a beer. The young man came over with his beer and said his name was Tony. He couldn't accept the beer because he was broke and would be unable to buy me one. I told him not to worry, to drink the beer, that one of these days I would be back and he could buy me a beer. We were on our second beer when a couple of prostitutes came in. They spoke to Tony before they went to the back and sat in a booth. Tony was generous enough to give me details about them, including prices. He said prettier girls would be in later if I wanted to wait. I told Tony I didn't have the time and I bade him farewell, and left to catch the bus to the airport. I never thought I would ever see this man again, but seven years later I would have another brief encounter with Tony for a second, and last, time.

I was very happy when we started packing for our return to Camp Lejeune. Three months in Vieques, Puerto Rico had been very hot, humid, and boring.

We made a very interesting stop on our way back to the States. We docked in Trinidad, British West Indies. I liked the island and the liberty so much that I requested a transfer to the marine guard detachment on the island. My request was later denied in a letter from Marine Corps headquarters.

It was good to be back in Camp Lejeune. It seemed we had been gone for ages.

I had spent a lot of time and work training these men as a team. I was extremely disappointed to see them transferred to the four corners of the corps. They would never operate as a team in combat. This was disheartening for me; for some reason I had thought they would remain together.

The following morning I received another shock. I was now facing 150 brand new burr-headed recruits, fresh out of boot camp. Standing in front of these men gives you sort of a helpless feeling, knowing you have a full complement of artillerymen or cannoneers of a 105-mm howitzer battery who have never laid eyes on an artillery gun. My work was cut out for me. I had been proud of my success in training the last bunch of recruits. I just hoped these men were equal to the others in their learning abilities. The training cycle began all over again, like a second hand in cards. Only this time it would be played with a joker in the uniform of a commissioned officer. This officer would jump on and trump just about anything I tried to advance or achieve while I was under his command.

I started training immediately, beginning with the basics. Each individual would be moved and trained in each position of the gun section until he knew the positions well enough to undertake a maneuver using live ammunition. This training had to be thorough, not only for future fire missions, but for their own safety in practice and maneuvers. I continued this training for over two weeks without ever seeing or hearing from the executive officer. This was unusual, but it wasn't my responsibility to keep track of the officers. I just assumed he was on leave, in school, or on some other regimental assignment.

In the operation of an artillery battery, the battery commander, besides being in command, works with the first sergeant in taking care of the clerical and office administration. The executive officer, with the battery gunnery sergeant, takes care of everything outside, such as the field duties and operations.

The gunnery sergeant is similar to a foreman on a ranch. He is the ramrod of the unit and acts as a buffer between the enlisted and the command officers. A good gunnery sergeant should have no problems in running the battery in the absence of the executive officer. Likewise the executive officer should be capable of running the battery in the absence of the gunnery sergeant.

By now we were a little over two weeks into the training schedule. I had the battery lined up on trucks, with the guns hitched on behind, waiting to move out in the Second Marine Division motor parade. While walking back to check on the troops and equipment, I noticed a lieutenant walking along beside the trucks, quizzing the men. I approached the officer and asked him why he was questioning my men. He gave me a dirty look and a very sorry, sarcastic answer. I didn't care for either one. I told the lieutenant if he wanted to question or inspect my men, I would have them fall out and into formation. The lieutenant said, "I might just do that, sergeant. I'm going to run your ass ragged while I'm the executive officer of this battery." This was news to me. I had not been informed of any change in the executive command. It was very plain from the expression on the lieutenant's face that he didn't like me. I certainly didn't care for his demeanor or attitude, but if he was going to be the new executive officer I would have to work with him.

I had worked with other officers and enlisted men that I was never overly fond of, but we generally had a mission to accomplish, and petty differences, attitudes, and mannerisms were something you had to accept or put aside. I was never naive enough to believe that everyone I worked with liked me, but I took pride in being tactful enough to work with anyone. I always tried to set a good example, not only for the men under my rank, but for the officers above. Up until now I never had any problems. The exception, of course, will always catch you by surprise. It can slowly creep up, strangle, and stifle your work, even before you are aware that it's taking place.

Trying to fulfill my duties as gunnery sergeant became very difficult as the training progressed. I was continually plagued with many petty and time-consuming orders. Nothing I tried to do seemed to stay on schedule. I had some of my orders countermanded, and at times I was given conflicting orders. Sometimes I was given orders that did nothing but cause a lot of trouble or resulted in lost time. An example would be holding an unscheduled inspection that kept the men waiting well after the other units had departed on liberty, then having me dismiss the men without completing the inspection.

I've served under many good officers. Some of them were very strict, but most of the ones I had known tried to be fair. I thought the lieutenant was just being a strict officer. This didn't bother me, because if I had an alternative I would prefer to serve under a strict officer. As time progressed it became evident that this had nothing to do with training. It was his way of putting obstacles in my path and harassing me. It was beginning to get under my skin, but I decided to ignore him and try to do the best I could under the circumstances. The lieutenant would send someone a mile down to the gun park with orders that he wanted to see me immediately. Once I walked the mile back to the office, he would give me some small and frivolous order that could have been taken care of at any time. It seemed impossible to make this officer understand our duties. I thought he didn't care, but the truth soon became known to me. He didn't know the first thing about artillery, and he didn't want to learn.

I spent two years serving as acting gunnery sergeant with no assistance from that executive officer. From the first day I met him he did nothing but hinder and harass me, thereby making my duties more difficult. Of course, during all of this time I never received one compliment. The lieutenant devoted most of his time to chewing me out about some minute detail that was never relevant to the training or operational objectives of the battery. I believe this was his way to camouflage his ignorance of artillery. Whenever he chewed me out, it was in a voice loud enough for the men to hear. More than once I asked him not to do this in front of the men, as it undermined what small authority I had over them.

I had two section chiefs, equal in rank with me. I wanted to teach them enough so that one of them could take over my duties when I was absent. It was quite obvious that the executive officer would not be able to do so. I could have taught the lieutenant a lot about artillery, but he was never around, and when he was present he was never in the mood for learning, especially from me. After observing the treatment I received from the lieutenant, neither of my two NCOs cared to act as gunnery sergeant.

As long as I had been in the marines, I always supported any superior officer. Regardless of my personal feelings, I would always tell the men we had the best CO and executive officer in the Marine Corps. To do otherwise would undermine the morale, discipline, and efficiency of the entire unit.

The lieutenant continuously plagued me with extra duty, and various other things to delay me leaving on liberty.

Many times when a field maneuver or an overnight exercise was scheduled, he would withhold this information from me until the morning of the exercise. Not having a telephone at home, I would have to leave without letting my wife know. At one time we were on maneuvers for two weeks, while my wife waited for me to come home every evening. This was done deliberately to create stumbling blocks and problems for me.

If I could remember every little thing I got chewed out for, I would

166

have a large volume. I'm only going to relate a few that I remember, just to give the reader some idea of what I experienced.

Many times the lieutenant would tell me to do something— something we had tried before and didn't work. If I tried to say something, or reason with him, he would become very upset. His voice would rise about two octaves, and in the higher-pitched voice he would almost scream, "I gave you a direct order, Sgt. Thomas, and I expect you to carry it out, or I'll have you on report." I heard that hundreds of times; I can still hear the echoes.

If I tried to explain or teach him something about artillery, he bolted or left abruptly without a word. I never understood why. Many times, officers would be put in a position to cross-train, knowing nothing about the position he was in.

In Korea I had officers join me on the front lines who knew nothing about being a forward observer. If they stayed with me for at least a month I could teach them how to fire and adjust artillery. I was always willing to teach an officer all I knew. I never had any secrets.

I never really knew why the lieutenant was down on me. Sometimes I thought it was jealousy. I had four rows of ribbons and he had only four ribbons. We each had the Silver Star.

One day when we were in a parade welcoming Gen. Chesty Puller as the new commander of the Second Marine Division, the general began to troop the line. When he got to the front of our unit, he turned and came straight to me. He congratulated me and asked where I had received the Silver Star. He bypassed a lieutenant who was also wearing the Silver Star ribbon. After this incident, things became practically unbearable. He not only made things rough on me, but on the battery as a whole. I was pressed to the point where I considered knocking his head off, regardless of the consequences. I believe he was aware of this, and this was why he would make sure he had an audience in case he needed witnesses to an assault on his person.

The executive officer's lack of knowledge of an artillery battery was always a stumbling block for me, always an obstacle to overcome.

I'll give the reader one in-depth look into the executive officer's command presence during a big and important general's inspection of our artillery battalion.

I knew I would not get nor did I expect any assistance or support from the lieutenant, so it would be up to me to get the battery set up and prepared for inspection. I had watched Sgt. William S. Prosser get a battery ready for a general's inspection years ago. This experience and the manual were all I had to work with in a very short period of time. Everything for inspection had to be laid out and displayed in a certain order. I had about three hours to go when I started the layout. I was in the final phase of lining up the guns, trucks, jeeps, and equipment using the aiming circle for perfect alignment. Everything was hustle and bustle. Time was getting short.

At about this time the executive officer made his appearance on the scene. The lieutenant came directly up to me and, in his high and most authoritative and commanding voice said, "Sgt. Thomas, I saw a couple of men with their boot laces hanging outside of their boots. I gave you a direct order, these laces are to be tucked inside the boots for this inspection."

I said, "Lieutenant, I'll take care of it," and continued with inspection readiness. I still had my hands full, trying to get the equipment laid out and lined up.

Here the lieutenant came again, and in his voice about two octaves higher screeched, "Sgt. Thomas, I gave you a direct order, and I expect you to carry it out. Do you understand me?"

I said, "Lieutenant, we don't have much time. You take over here, and I'll obey your order right now."

He bounced off like a bantam rooster with his feathers ruffled. Once the equipment was ready for inspection, I had the troops fall into formation. These men had been busy getting their guns ready; they hadn't had time to check their dress. Now in formation, I asked all of them to bend over and tuck the laces inside the boots. A very simple procedure. The lieutenant arrived just in time to stand in front of the battery as the general made his inspection. We passed the inspection with flying colors.

Some of the sergeants told me that I should go and complain to the captain about the lieutenant, but I was not a crybaby, and I didn't intend to go crying to the captain. The lieutenant was irritating, obnoxious, and probably spent much of his time creating problems for me. None of the things he had done to date were any worse than what most of us had endured during boot camp training. I knew the lieutenant despised me for some reason; this was evident in his not returning my salute properly. He had other ways of displaying his dislike for me. I reciprocated mostly by being insolent. This became a normal reaction for me. It was the only act of defiance that I could think of that wouldn't get me into trouble, and still let the lieutenant know what I thought of some of his trivial orders.

We were scheduled for a six-month Mediterranean cruise. This cruise was with the Sixth Fleet, and would be a floating readiness force that could be dispatched to any trouble spot in the region on a very short notice.

While preparing for the cruise, a tech or gunnery sergeant was transferred into our battery. He would be the gunnery sergeant for the battery during the Mediterranean cruise. I was very happy to hear this; it would remove me from direct contact with the executive officer. I looked forward to this respite.

One morning I was summoned to the CO's office where all of the staff officers were assembled for a briefing prior to our departure on the Mediterranean cruise. I was asked to stand at ease. The captain informed me that the other gunnery sergeant did not work out. He was being transferred to another unit. Time was short, and the captain didn't have the time to request and have another gunnery sergeant transferred into the battery. He asked if I would be willing to act as a gunnery sergeant on the Mediterranean cruise. I didn't relish being gunnery sergeant with the present executive officer, so I decided to speak up about this problem. I explained to the CO that performing the duties as gunnery sergeant was hampered and made very difficult by the executive officer continually berating and chewing me out in front of the men. I also mentioned that the lieutenant continually plagued me with small and childish orders while I was busy with the operation and training of the battery. The captain confronted the lieutenant about this, and suddenly the lieutenant was all peaches and cream. He apologized to me and said he didn't realize his orders were so difficult to carry out. I was taken by surprise. The lieutenant called me gunny for the first time. The captain said, "Sgt. Thomas has been doing an excellent job as gunnery sergeant, so it's all settled. He will be the gunnery sergeant for the Mediterranean cruise." I was dismissed, but still uneasy about the executive officer. Peaches and cream were strictly out of character for him. I still considered him despicable and a very sorry excuse for a commissioned marine officer.

In January 1954 our convoy departed Camp Lejeune for the port city of Morehead City, North Carolina. The men in G-3-10 embarked aboard three ships. About a fourth of the battery and I embarked aboard the USS *Whitemarsh* with all the guns and equipment. My CO executive officer and the first sergeant would also be aboard this ship.

Once underway, I was informed that I was the senior NCO on board, and I would be the acting gunnery sergeant for all the troops on board. This included tanks, shore party, and some engineers. Shore patrol, guard, and security of the ship would be my responsibility. This was an added workload, but I had no choice in this assignment.

This ship was an LSD, or landing ship dock. The center of the well deck could be flooded, and once the ramp was lowered the amphibious vehicles could enter or leave by floating in and out of the rear of the ship.

While in the Mediterranean with the Sixth Fleet, G-3-10 would be the artillery support for the First Battalion, Eighth Marines, commanded by Lt. Col. V. H. Broertjes.

Since the meeting with the staff officers, I had very few problems with the executive officer. The only time he would call me gunny was when the CO was present. His dislike for me was still quite evident in his mannerisms.

Our guns, trucks, and jeeps were loaded on the top rear deck or afterdeck of the ship. I worried about the steel cables or slings used to hoist the equipment aboard. Some of these cable splices had become frayed. I had sent some of them to our ordinance for repair. When I got them back from ordinance they looked real neat. All the splices were taped. I removed the tape and found the frayed ends had been cut off and the splices taped over. This weakened the joint. Ordinance told me they didn't know how to splice cables. One evening while cleaning the guns on the afterdeck, I mentioned to the lieutenant that this would be a good opportunity to get the navy cable splicers to repair the slings. The lieutenant exploded and, in front of the men, he yelled, "Sgt. Thomas, you take care of the men and their training. The equipment is the responsibility of the executive officer, so it's my responsibility, not yours. Do you understand that?"

All I could say was, "Yes sir."

Later in the Mediterranean cruise while preparing to make a landing on the island of Sardinia, the sling on the captain's expensive radio jeep gave way while the jeep was high in the air over the well deck. It came crashing down onto the well deck, landing upside down; one wheel came off, and the radio was crushed. I couldn't let that pass. I said, "Lieutenant, that's your responsibility," and departed to locate my motor sergeant.

I was in my compartment telling my sergeant how to remove the jeep, so we could get on with our maneuvers, when the lieutenant came in with a full head of steam. He was very red in the face, and in his high screeching voice said, "Sgt. Thomas, get up there right now, and get that jeep out of the well deck." I told the lieutenant that my motor sergeant would take some men up and remove the jeep. He extended his vocal chords and practically screamed, "I gave you a direct order. I told you to clean it up, not the motor sergeant."

"I'm the gunnery sergeant, lieutenant. I don't have to do it. I'll have it done the proper way," I responded.

He turned and bolted for the stairway. He didn't put me on report as I expected, and I never heard any more about this, but I knew he wouldn't give up. You could bet your life he would find other and better ways to retaliate.

We had been very lucky that no one was killed or injured in this accident. The jeep had been picked up by the crane and moved across the deck of the ship, over the heads of many of the troops waiting to board landing crafts, before it broke loose and landed in the well deck.

There was always something to keep you busy aboard ship. I was always looking for something else to do, to keep my mind busy, even in my spare time. If I let myself be idle, many and varied thoughts would cross my mind. Life aboard this ship was different from the other ships I had been on. For one thing, we were not going into combat, but floating around waiting for something to happen. To me this was boring. I needed a war to fight. Without a war to fight, my hostilities were directed to other things. I had no remedy, but remaining in the marines with no war to fight was a very depressing thought. I was losing my desire to train men, only to have them discharged to civilian life. After this Mediterranean cruise, I would request a transfer. I could see no need to remain in the fleet marine force during peacetime. Being continually subjugated by a substandard lieutenant was not my idea of a career. There had to be something better. One thing was certain, I could never support or raise a family while in the marines.

Our first stop in the Mediterranean was the port city of Oran. The city, looking from the bay, appeared to be very dirty. Oran is a very old city, once ruled and governed by the ancient Romans. It was raided by the Barbary pirates in the seventeenth century, and was the seat of the French government in World War II. It had the overall look of a slum city. Nonetheless, it would be nice to get off the ship and look the town over.

Our CO's wife met the captain in each port. The CO debarked and left on the captain's skiff as soon as we dropped anchor. The captain made this a nice vacation or second honeymoon. He departed once we dropped anchor, and only returned before we raised anchor to get underway. This was his routine for the entire six months in the Mediterranean. We had a good captain, and I was happy to see him take advantage of the Mediterranean cruise for a second honeymoon.

One thing I didn't like about this arrangement was that I would be at the mercy of the executive officer, who would be in command of the marines on board this ship. The lieutenant, over the next six months, would continue to use any excuse he could think of to deprive me of liberty ashore. He took the first opportunity to exercise his authority by depriving me of my liberty in the first port. He caught me just as I was about to depart the ship, and said the present roster I was using would be unacceptable. He wanted a new one made before I went ashore. This was an order that was impossible to carry out, because of the alphabetical listing of all the marines on board. It was a coincidence, but all but a couple of my sergeants ended up on the third list. The main roster was divided into three equal parts: One-third would be liberty, one-third standby, and the other list would be the duty personnel. In order to alleviate the shortage of NCOs, I made three more lists using only the NCOs. This would insure equal duty for all of the NCOs. I went to the first sergeant with that list and obtained his approval. That was the only way I could think of to be fair to everyone. At evening mess I ran into the first sergeant, who asked why I wasn't on liberty. I explained what the lieutenant had told me to do before going on liberty. This was the first time I had ever heard the first sergeant swear. He said, "I'll take care of it. You go ahead and catch the next liberty boat." I continued using my roster and never heard any more about this from the executive officer. However, the lieutenant was very crafty in coming up with other excuses to block me from going on liberty in the various ports we visited while in the Mediterranean.

Since the executive officer was set on depriving me of liberty, it was up to me to improvise a method to circumvent this injustice. The only time the lieutenant looked for me was on the day I had liberty. On the other days, I never saw or heard from him.

While still in the port of Oran, my duty was to furnish two shore patrolmen to accompany a desert tour of marines going to visit the headquarters of the French Foreign Legion. Since it was my duty day, I assigned myself as one of the shore patrolmen.

The Legion headquarters was located about sixty miles from Oran in the small town of Sedi Bel-Abbes. The bus trip out to this town was long, hot, and very dusty. We passed many people in Arab dress riding or leading camels who were burdened with a big load. Many of the people were just walking, some with small children, miles from any sign of habitation.

Visiting the Legion headquarters proved to be very interesting. Cpl. Leon Wasielewski, one of our men, acted as interpreter while a Polish officer in the Legion conducted the tour through the compound. In one stucco building there were hundreds of pictures of legionnaires of every nationality who had been decorated. Many had been killed in the various actions of the Legion over the years. Many Americans were listed, but most of these men had used an alias, and were listed as Bill Smith, American, or Bob Jones. The real names of these men who had gallantly distinguished themselves in the various battles of the Legion will never be known.

In another stucco building was a weapons museum housing hundreds of weird enemy weapons of every description, old and new, that had been captured in many of the desert skirmishes since the Legion was organized.

The Legion had just recently withdrawn from the fighting in Indo-china, where it had suffered many casualties. There were many wheelchairs in and around the compound. Any legionnaire who was disabled could live in the compound. This was the only disability retirement available at this time.

While on the tour, six dirty, unshaven men passed us. They were being guarded by three legionnaires. I asked my interpreter about these men, and learned that they were in the brig. Most of the men in the brig were generally sentenced from six months to a year. During this time they were only allowed water to drink. Part of the punishment was not being able to shave, bathe, or wash your clothing. This appeared to be true, judging from their unkempt appearance.

I talked to the Legion officer about joining the Legion. The least amount of time you can sign up for is six years. The pay then was twenty dollars a month. If you can speak two languages fluently, you can become an officer. The pay of a lieutenant then was forty dollars a month. If you became disabled, you received free room and board at the Legion head-quarters for life (not enough incentive for me to sign up). We returned late that evening to the port of Oran, and the USS *Whitemarsh*. This was the only liberty I made in Algeria.

It would be nice to write a travelogue of all the ports we visited, but I was stopped many times prior to departing the ship by the executive officer who would have some complaint, causing me to have to change clothes and delaying me from going on liberty until hours later. I missed going ashore in two ports, and missed many hours of liberty because of something the lieutenant would find necessary for me to do prior to leaving the ship. I didn't dare complain when the captain returned to the ship, because as things stood, I was able to make some liberties. If I said anything, I'm sure the lieutenant would have made my predicament much worse than it was. By missing liberties, I was able to send over three hundred dollars home to my wife, so it wasn't all that bad.

I was proud of the operation of the battery during the maneuvers on Sardinia. We had trained the men well, but a giant storm cut our man-euvers short. When we left Sardinia, it was a muddy quagmire.

In Naples, Italy liberty went at twelve noon. I was delayed again and didn't get to leave the ship until six that evening. While in this port, one of my section chiefs failed to return from liberty. I passed the word to my shore patrol to be on the lookout for this sergeant. One of the shore patrolmen reported seeing this man in a café between twelve midnight and one o'clock. The following night I assigned myself to shore patrol duty. Liberty ends for the enlisted at twelve midnight. Shore patrol stays until about one in the morning to round up strays, or anyone missing the last liberty boat.

At twelve midnight I entered this small café on the edge of town. My section chief was sitting with a beautiful blonde girl. He asked if I had come to arrest him. I said, "No, I just wanted to talk and find out if you are coming back." He said that no, he was going to stay in this city. Nothing I could say would change his mind. He made the remark that all of the years I had spent in the marines counted for nothing. He didn't like the way I was being treated and remarked that if this was all he had to look forward to in the marines, it was time to bail out. I was short on time, so I wished him good luck and departed. Back aboard ship I reported him as being AWOL. I never told anyone about our meeting. I still had hopes that he would come back on his own, but I never heard from him again.

I made two memorable liberties, one in Malaga, Spain, and the other in Nice, France. While in France, a tour was leaving for Frankfurt, Germany. I was supposed to furnish two shore patrolmen to accompany that tour. I assigned myself as one of the shore patrolmen. This time I was not so lucky. The executive officer found out about this, and I was ordered to remain aboard ship while I was in a duty status. After this encounter I was unable to use this tactic as a means of going ashore.

The crew of the USS *Whitemarsh* did their best to make our cruise enjoyable, but after six long months aboard this ship in the Mediterranean it was a happy day when it docked in Morehead City, North Carolina.

It was good to be back in Camp Lejeune. It took over a week to get all of the equipment cleaned and inspected. Once this was completed, I decided to ask for leave to visit my wife and son in San Antonio, and my family in Dallas. The previous year, I was refused the twenty days requested. The executive officer would only allow me ten days furlough. It took three days on the road to reach Dallas, and another day to get to San Antonio. I like to allow four days to return to the base in case of car trouble. I only spent one day with my wife and son, and one day with the folks, before starting the long journey back to Camp Lejeune, North Carolina. This time it was no different; the lieutenant would only give me ten days leave. I explained to the lieutenant that I needed more than ten days, and requested he give me at least fifteen days, but he refused. Ten days would be all the leave he would let me have.

My marriage was already in trouble, so I decided to go straight to San Antonio and skip a visit with my family. This would give me more time with the wife and son.

I departed on leave with codeine tablets for my back, which hurt while driving, and some No-Doz tablets. To have more time at home I was determined to drive the distance in the shortest time possible.

It took one day to reach Macon, Georgia, before my car broke down. It took three days to repair the car. Because of the loss of time, money, and the uncertainty of the car, I decided against trying to make it to San Antonio. On the way back to Camp Lejeune I decided to spend a couple of days of my leave in a beach resort located in Myrtle Beach, South Carolina. I returned to the base to pick up my mail and bathing suit. The mail was bad news; the wife was filing for divorce and there was nothing I could do about it. I was depressed and angry because I couldn't get enough time to attend to my personal problems. Maybe a couple of days in the sun in Myrtle Beach would help me to reason something out. There had to be some way to circumvent one man from destroying my life. No use going to the captain. He was already aware of the problem, but he did nothing to help. I wasn't about to approach him again.

Deserting had been a thought, but it would not be a solution. What alternative was left? My mind was in turmoil as I left the staff NCO quarters and started for the parking lot.

The staff NCO quarters are quite a distance from the battalion head-quarters, so I was very surprised when I found the executive officer blocking my path to the parking lot. He said, "Sgt. Thomas, we are getting ready to load up for maneuvers in Fort Bragg. Go to the battery storeroom and make sure the trucks are loaded properly." I told the lieutenant that I still had a few days left on leave, and I gave him the name of one of my section chiefs who could load the trucks properly. I continued to walk toward the parking lot. The lieutenant started backing up, and then started to yell that he was giving me a direct order, and I had better obey his command. I said, "Lieutenant, I've had enough of you. I"m leaving, and I don't think you are man enough to stop me, so get out of my way." He stepped aside, and I continued to the car with his high-pitched voice screaming behind me about what he was going to do. I got in the car and departed for the resort town of Myrtle Beach, South Carolina.

My dreams of lying on the beach were shattered when a rainstorm hit Myrtle Beach in the late afternoon. I drove back to Wilmington, North Carolina, and visited a couple of clubs. With nothing else to do, I decided to go back to Camp Lejeune. If the lieutenant was bluffing, as he generally did, I would check in early from leave and take over my duties as usual. This time he wasn't bluffing. I had been put on report for direct dis-obedience to a lawful order. I was ordered to report to the battalion commander.

My building hate, combined with an uncontrollable rage, mixed with a fiery temper is an explosive combination. When combined they only need a small catalyst or primer to ignite a conflagration that cannot be imagined or controlled by the most vivid imagining possible. This is a lurking monster that pounces on your sanity like a driven demon, set to destroy your mind and soul when you least expect it. There is no warning. When you reach the pressure threshold, you have crossed the boundary of rationality, and the nightmares from the secret chambers of your mind are released on the unsuspecting at a most vulnerable time. The mind is like a miniature solar system, with millions of unexplored crevasses and a transportation system faster than the speed of light, that can bounce a thought from the remotest planetary plane of the mind to the present. A million triggers are set and can be released spontaneously, exploding brain waves and the normal thought patterns into oblivion.

I did my best to try and rationalize and analyze the culmination of a situation that, if remained unresolved, would eventually destroy me and possibly those that oppose me. Now at last I could lay this on the battalion commander. Maybe after hearing my predicament he could offer a solu-tion. I had never met this colonel, so I knew nothing about him. Maybe now I could offer my problems to a higher echelon. I intended to ask the colonel for emergency leave or a transfer. I thought perhaps the colonel would understand my plight. What a surprise I had. He would accept no criticism of any officer in his command. Once I mentioned a personality conflict the colonel went wild. I was told there was no such thing as a personality difference between an officer and an enlisted man. He word was final. I was reprimanded severely. He would hear nothing else that I had to say. Even in a court-martial the defendant is offered to allow mitigating and extenuating circumstances in his defense. This was the end of the line. I had had enough; I was leaving the Marine Corps.

As I left battalion headquarters on the way to the parking lot, I saw my executive officer talking to another lieutenant. As I came closer the lieutenant departed, as he passed me. I gave him a snappy salute. I passed within ten feet of the executive officer; I look straight at him and did not salute. I waited for one word, anything. I was prepared for a fight, even if it meant losing my stripes and doing brig time. Once I passed the executive officer, I passed very slowly, hoping he would catch up and say or do something to provoke me. I was shaking with rage, hoping for a confrontation. I didn't look back, but went straight to the car, opened the door, and sat down. There was no seat. My body melted into the darkness of a black hole where the seat had been. It swallowed me up.

The noise from the train engines created a deafening roar. They made a thunderous noise as they banged into the boxcar. Inside it was so dark. You couldn't see your hand in front of your face. I could hear the loud whirring and clickity clack of the train wheels as they gained momentum. The speed increased in direct proportion to the increasing noise of the wheels. The noise increased to a level that was beginning to hurt my ears. I felt around in the boxcar, trying to find something to hold onto; there was nothing. I fell and started falling down a spiral tunnel. I was not falling through the spiral, but falling around the outside of the spiral at a tremendous rate of speed. I kept falling and falling; it was taking forever. At last the spirals was beginning to get smaller. Finally I hit the bottom of the abyss. At last it was over. I could get some rest. The crescendo of the exploding engines startled me as they started again. Oh, no! I began to fall up and out of the pit. I started at the narrow bottom, falling upward, and traveling at an unbelievable rate of speed around the spiral as the huge rings increased in size. It seemed to go on forever, but at long last, as the spirals became larger in diameter, the speed began decreasing in proportion. The spirals became so large, and the speed so slow, that it was taking forever to make one revolution.

At long last I was back in the darkness of the boxcar. The deafening roar of the train engines abated; all you could hear was the steam being released from the engine with a loud hissing noise. I found the bulkhead of the boxcar and sat up. The boxcar began closing in; it was getting smaller. It was still dark; I couldn't see but could feel or sense the space decreasing. I was getting scared. I had to find a way out before I was crushed to death. A noise, a noise in the darkness. I was not alone. Something was lurking in the darkness very close to me. Was this human or animal? I lost my weapon; I had only my hands to defend myself. Light began to emerge slowly as a cloud lifted. I found myself back in my car with a strong grip on the steering wheel. I was still in the parking lot. I do not know how long I had been sitting in the parking lot.

Once my senses returned I wondered if I was cracking up. I had seen many men with combat fatigue. Maybe it was my time. I drove to the Camp Lejeune naval hospital and asked to be checked in. I was informed that I would need to get an admittance slip from my battalion doctor. I went back to battalion sick bay and asked to see the doctor. I was told to come back at sick call the following morning. I said I was deserting the Marine Corps. If I couldn't see the doctor I was leaving right then and there. I saw the doctor and received the admittance slip I needed to check into the hospital.

I'm not sure what I expected as I checked into the psychiatric ward of the naval hospital. I was assigned to a bunk in a ward that contained about twenty bunks. Only six were occupied. One man had tried to commit suicide; one was being treated for alcoholism. None were raving maniacs.

We were up at reveille every morning, made our bunks, and cleaned the ward. Inspection went every morning at nine o'clock. The inspection was carried out by the doctor, a navy commander, who asked each man how he was doing. This was the same routine every morning.

It was two days before I was summoned to the doctor's office. He wanted to know what problems caused me to be in the psychiatric ward of the hospital. He also wanted to know if I drank or used narcotics. He recorded the medical history of my family back to my grandparents. As for my problems, I told the doctor that the last time I had made a statement about personality differences in front of a commissioned officer, I was severely reprimanded, so I was not going to pursue that problem. I did relate and explain that I was becoming very depressed with peacetime service. I was used to fighting in a war or conflict, and without this or something else as an objective, my life in the Marine Corps seemed wasted or useless. As a result I felt that my usefulness to the marines had come to an end. I did not mention the nightmarish experience that occurred while sitting in my car. When I was asked if I was taking any kind of narcotic, I answered "No," but later it occurred to me that I had been taking codeine and No-Doz tablets. I had never associated pain pills with narcotics. Codeine for my back, and No-Doz for keeping me awake while driving, suddenly struck me as an unlikely combination. Being upset, I had skipped eating regularly, and the night before the frightening experience, I had had too many drinks to count. This, I'm sure, was the side effect from combining pills and alcohol.

Over the next two weeks I underwent various psychiatric testing, including the inkblot test. During that time I was allowed to go on liberty after the morning inspection. I was warned not to return to my unit for any reason, and I was to be back in the barracks by five in the evening.

One day during the fourth week, the doctor called me into his office. He gave me the results of the psychiatric evaluation which indicated that I had a certain amount of depression. The doctor said, "Sgt. Thomas, I did something I have never done for any patient under my care. I went out of my way to spend most of a day talking to the men in your unit. I did this as a courtesy to you, in view of the many combat ribbons you are wearing and your length of service in the marines. I found out some interesting things about you. You have an outstanding record in the Marine Corps. Every fitness report that has ever been submitted on you since you became an NCO has been outstanding. I also learned of the ongoing conflict between you and the executive officer. If this has been your biggest problem, it will be resolved shortly. He is being discharged within the next thirty days. Your commanding officer informed me that your promotion to the rank of gunnery sergeant is in battalion headquarters. If you want to go back, receive your promotion, and resume your duties as gunnery sergeant, I'll keep you here until the executive officer has been discharged. Now, in view of your outstanding service in the marines, you tell me you want out of the Marine Corps, and I'll get you out. I want your answer tomorrow morning. Once you give me your decision there can be no turning back."

The following morning I had another chat with the doctor. I told him all about the executive officer and my feelings about him, but that this is a position that will always be occupied by a first or second lieutenant. I said, "The one leaving might be a lot better than the next one. I never thought I'd have any trouble working with any officer in the Marine Corps, but I've been in the marines much too long. I know how a battery should be run, and I become very upset when an inexperienced officer tries new experiments with my men and equipment. It's not hard to make a choice. I'll never be intimidated or dominated by another inexperienced officer; I just don't care to take that gamble. I hate to do it but, with no war to fight, I'd like to leave the Marine Corps as soon as possible."

It took two more weeks before I was able to leave. The doctor shook my hand and wished me luck in civilian life. I thanked the doctor for all he had done for me while in the hospital. He was a very considerate officer. I could leave the following morning, but I was given a strict warning. I would not under any circumstances be allowed to return to the battalion area or the NCO barracks prior to leaving the base. My pay and discharge would be mailed to me.

The following morning I departed the naval hospital. On my way off the base, I came to a crossroad. The one going straight ahead for a few blocks would lead to the battalion parking lot. By turning left I would be on the main road leading to the main gate. I sat at the intersection with mixed emotions. I had men I wanted to say goodbye to, and I needed to pick up some personal possessions in my locker in the NCO headquarters.

I sat at the crossroads for quite some time trying to make up my mind. My marine unit had been like a family to me. I hated to leave without a farewell, because I knew I would never be back on this base. This time my leave would be permanent.

I promised the doctor I would not return to the battalion area. Since he went out of his way to aid me, I was compelled to honor his order and request, regardless of my personal feelings. A car pulled up behind me. With mixed feelings, I turned left and was on the main road leading to the main gate of Camp Lejeune, North Carolina.

I had made my decision. Good or bad, you make it and hope that in the future, whether you are successful or not, al least you have the control of your life in your own hands.

I had just completed my last and final campaign in the Marine Corps, and I could truthfully say I was leaving with plenty of mixed feelings. It wasn't the ending I had hoped for, but an ending nevertheless. Another road lay ahead, another battle, another beachhead. I doubt if I'll ever stop fighting. In time perhaps the memories of the battles will subside, but the memories of all the men who served beside me under the most adverse circumstances will never fade as long as I live.

I've had the special honor and privilege of serving with the best marines who ever laced up their leggings, put on their boondockers, or packed a sea bag for combat duty beyond the sea.

I felt sad as I passed through the main gate of Camp Lejeune for the last time. I gave a final salute as I departed, closing out twelve years of my life. I left with the full knowledge that I had tried to be the best marine sergeant in the Marine Corps. It was like leaving home for the first time. I was departing with a heavy heart and an emptiness that I had not experienced before. I would miss the marines, but there comes a time when you have to make a choice, not knowing if it's for better or worse. I can say honestly that during all of the years I spent in the Marine Corps, I never met a marine I didn't like, except one.

After I had cleared the main gate on my way into Jacksonville, words started flowing through my mind, and each time this happened, I pulled off to the side of the road and wrote them down. When I had arrived in Jacksonville, I had finished what I called "My Farewell to the Corps."

My Farewell to the Corps

I'm proud of the men who have fought, and died.
Heroic men, who have served by my side.
For every crest, there must be a fall.
I've served with honor, I heeded the call.
The wars are passed, but the enemy lurks.
In the peacetime duty, of an officer's quirk.
Many times I've danced, when sirs you obey.
None as lively, as the boondocker ballet.
A call to arms was an honorable chore.
The call is over, I'm leaving the corps.
As I pass through the gate, with a final farewell.
A salute is in order, I have broken the spell.
I'll remember the corps, I served it with pride.
The fears we all suffered, we took them in stride.
I leave with regrets, a distinguished past.
The bonds are severed, from an officer's lambaste.
On the long road home, a shiny new light.
Peacetime survival, I have fought my last fight.

S/Sgt Melvin H. Thomas
USMC 820698

Epilogue

I received my honorable discharge in September 1955, just short of twelve years in the corps.

I found my marriage to be beyond repair, so I ended up with two divorces—one from the marines, and one from my wife. It's possible things could have been different if I had been given a furlough to go home when I had made the request, but I have no way of knowing this.

I wasn't about to go back to Dallas and waste my time with the unemployment office, so I decided to remain in San Antonio. At thirty-one years old, I had to have a new career and start all over again. Radio broadcasting was out and TV was the up-and-coming thing, but I knew of no schools teaching this so I decided to pursue my training in electronics. I completed a four-year school studying electronics and radar. To support myself while attending school, I worked nights and weekends at the Lloyd Ball Texaco station on North Saint Mary's Street. After completing my four-year study, I entered an advanced school to learn the B-52 bombing system. On completion of this training, I started working as a systems analyst on the B-52 bombing system at Kelly Air Force Base. This was the old K-system, and we were told while in training that this system was obsolete. It was due to be replaced at any time with the newer black box system. I had put the Marine Corps behind and spent most of my time studying and moving ahead with a new career.

While drinking a cup of coffee in a drug store in North San Antonio, I chanced to run into James Didier. He was one of my wiremen in Korea. He was attending Saint Anthony's Academy, studying to be a Catholic priest. I had another visit with him one year later before losing all contact with him. He was one good and dedicated marine. I sure hope he achieved his goal.

I remarried—a nineteen-year-old girl from Oasis, Utah—and ended up having five children over a period of five years. This included twins.

With my position as a systems analyst on the B-52 bombing system coming to an end, I decided to change over to the Federal Aviation Agency. I went to the FAA center and took some tests and examinations. I was selected for a position at the international airport in San Juan, Puerto Rico, and entered on duty at the San Juan International Airport in 1959.

Housing in Puerto Rico was very scarce. It took me three months to find and lease a house. I was lucky to have found this house near the airport. The twins were born in Hollywood, California, and my wife waited for a doctor to release them for air travel. All I had was a picture of them when they were born.

I was working with a number of Puerto Rican men who wanted to take me on a grand tour of old San Juan and a number of other small towns. I didn't have a car, so on one Saturday evening they picked me up at the house and took me to old San Juan to start the tour. After this they drove around until I was completely lost, but it didn't seem to matter. They drove by their homes to show me where they lived. We stopped in the town of Santurce and ate some Puerto Rican treats called mafungos. These are pork balls that taste like pork mixed with cornbread stuffing. They are rather tasty. Our next stop was a dance hall on the outskirts of Santurce. The dance hall had a good band and was crowded with a colorful mixture of the local population. I was enjoying myself, and about eleven o'clock they told me they had to run home, but they would be back to pick me up before closing time. I didn't mind; it gave me something to do. I was living in a house with no furniture, so I wasn't really in a hurry to return home on a weekend.

The dance hall closed, and I waited and waited, but my buddies didn't return. I was stranded all by myself, out in the middle of a field and very close to a stinking bay. It was about three miles from the main part of Santurce, so I decided to walk into town and catch a bus to the airport. The buses to the airport ran all night. I could see the tall buildings in the distance, so I decided to take a shortcut along the bay that led toward the town. I was already well into the area before realizing it was a slum area. There were no lights except those that were reflected from the bay. Nothing but shanties and lean-tos. The drainage ditch beside the road smelled like, and was probably used as, a sewer. The residents of this community would take advantage of anyone suspected of having a dime. I tried to tiptoe through this area. Luckily, I didn't meet or see anyone, and there were no barking dogs. I thought I would come out very close to the Zombie Lounge. This would only be a couple of blocks from the bus stop. I remembered the time I had waited for a bus and had dropped in at the Zombie and had bought a few beers for a curly-headed Puerto Rican boy named Tony. While I walked I could see the lights from the main avenue in the distance.

I was about three blocks away when I saw four figures turn off the main drag and start toward me on this street. At this time of the early morning it spelled trouble. Every time I crossed over to the other side, they, too, crossed. I tried walking in the middle of the street. They then split up and had two men on each side. I expected the worst. I was going to try and put two down quickly, and then take my chances with the other two. Just as they stepped off the curb toward me, I saw the curly hair from the back lighting of the main avenue. I said, "Hi, Tony, how are you?"

"My friend," he said. "I remember you; long time ago you buy me beer."

I felt very relieved and offered to buy them a beer, but Tony said the bars were closed for the night. I gave him ten dollars and told him to buy everyone a beer when the bars reopened. I told him I was working in Puerto Rico and would probably see him again. They continued on down the street toward the slum area, and I caught my bus back to the airport counting my blessings and thanking my guardian angel for working overtime.

Our daughter was born in the Santurce hospital in 1960. This made five children; it was enough. Five proved to be a handful; we weren't about to have any more.

When the war in Vietnam started, I felt like a football player sidelined on the bench. I felt left out, not being able to go. If I had known this war was going to happen, I would have probably stayed in the Marine Corps. If I had fought in that war, I'm quite certain I would have been killed. I would have volunteered to be back on the front lines calling artillery fire on the enemy. I had a brother who joined the marines about this time. He came to visit us in Puerto Rico. For some reason, because of his duties, he was never sent to Vietnam. I submitted a bid to the FAA on one communications position at the Saigon airport and thought I had a good chance of getting it, because being in a combat zone didn't scare me. For some reason I never heard any more about it.

I spent over two years in Puerto Rico, maintaining the microwave link between the airports of St. Thomas, St. Croix, and San Juan. I also maintained the remote air-ground facility for San Juan. Both of these facilities were located on top of the mountain El Yunque. This mountain overlooked the Roosevelt Roads naval station.

I received a transfer from Puerto Rico to Miami, Florida, and went to work at the overseas transmitter site, located just north of Homestead, Florida. Most of my work for the FAA would be very boring to anyone not interested in electronics. I generally worked or manned a remote facility unsupervised. This was a very responsible position because my equipment was involved in air safety. I was required to make many trips back to the aeronautical center in Oklahoma City for attending school on various equipment. At one time I had to take a supervised four-hour examination on advanced engineering mathematics. I had to pass this examination or lose my GS rating with the Federal Aviation Administration. I passed the test with little trouble, expending more sweat than necessary.

In 1991 I heard about and joined the Second and First Marine Division Associations. I was able to contract many of the men I had known back in the Pacific and Korea. It was great to hear from these men, and at the same time disappointing to hear of some of the men who had passed on.

I've lived my life striving to be the best in whatever I set out to do. At times I have fallen short, but I never lost anything by trying. I always believed that things happen for the best. I still believe this.

I was divorced in 1976, after twenty-three years of marriage and five wonderful children. The most rewarding and demanding thing I ever accomplished in civilian life was raising my kids. Besides being a blessing, it was also a twenty-four hour job trying to study and keep up with my career and, at the same time, trying to spend some time at home with my family. I do believe I had a better family life being a civilian than I would ever have had by staying in the service. Two years after my divorce I met and married a little Korean girl. As of now we have been married for seventeen years. She was a Korean orphan and she was never able to have any children.

I could list pages of top marines in my mind that I've had the honor and privilege of serving with during the Pacific and Korean wars. I know that many of their records and accomplishments far exceed mine. If they ever decide to write a book, I'll be the first to read it. I wish to express my appreciation to the many men and women in whose presence I've had the honor and pleasure of sharing, working, serving, and fighting during my thirty-four years of government service.

I sincerely hope you have enjoyed this narrative, and I also hope anyone who reads my autobiography will at some time in his or her life benefit from my experiences.

I can only write my own story as I know it. It must be remembered that at times I was on the low end of the totem pole, and I have used some of the information that filtered down to me. Some of this information might be considered hearsay, but I can only report what was communicated to me at the time.

In summary, I have never criticized any war. It never really mattered to me what nations were right or wrong. I was never qualified to judge any but my own, but I was always qualified to serve and fight. This was what I had succeeded in doing. My self-esteem has always benefited because I was never a sick bay soldier, and I very seldom complained. I always strived to do my best in any environment and I close with no apologies to anyone for anything I ever did in serving and performing my duties during war or peacetime.

Earthly Peace

Savage storms, brave warriors caught.
Destructive clouds, battles fought.
Radiant fields, where spirits soar.
Defeated nations, senseless gore.
Lightning dust, forever stains.
Barren oceans, muted plains.
Stormy echoes, o'er earthly scars.
Behold!, beyond the storm, the stars.

Melvin H. Thomas